For Auriol, Ben and Theo

Contents

Introduction

As I was finishing this book, some scientists had begun a campaign to warn people about a low-budget American film about life, the universe and everything. The film was called *What the Bleep Do We Know?*, and used a mix of documentary and drama to convey the message that there's a lot about our universe we don't understand.

How could any scientist argue with that? They can be a bit arrogant sometimes, but not even the most gung-ho physicist would claim to possess all the keys to the cosmos. Even so, some of them felt moved to issue public warnings about the film, describing it in terms ranging from the peremptory ("atrocious") through to the darkly disturbing ("It is a very dangerous piece of work").

Clearly, this was a must-see movie. Yet for the first twenty minutes or so, I couldn't see what all the fuss was about. Various scientists made pretty innocuous statements about how new discoveries were revealing the universe to be far stranger than anyone had expected.

It was when the film moved on to describe some of these discoveries that I began to see what the fuss was really about – discoveries like the ability of water molecules to be affected by thought. I had heard rumours of this before: of how a Japanese researcher had shown that the very shape of a water molecule could be radically altered just by the thoughts of those around it. Yet when the movie described this astonishing discovery, all it gave by way of evidence was some pictures of ice crystals looking nice after being blessed by a Zen monk, and looking nasty after being exposed to someone in a bad mood.

Now, lots of people do find this kind of evidence pretty compelling. It's immediate, clear, and apt to prompt wide smiles and sentiments like "Hey, man, far out". Among scientists, though, the typical response has been: "Give me a break". I know what they mean. Sure, the idea that water is affected by thoughts is an amazing claim with huge implications. With no obvious explanation, it also raises the possibility of radical new forces at work in the cosmos. But before we get too carried away, it might be an idea to have some decent evidence that the effect is real. And pretty photos of crystals, nice as they undoubtedly are, just don't cut it.

Many other similarly bizarre claims popped up in the film, along with similarly feeble evidence backing them up. But while I can understand why so many scientists became indignant over the film, I think they really missed the point. The real problem is that its claims weren't bizarre *enough*.

Water molecules affected by weird forces? Forget Zen monks, try this: water molecules owe their properties to a form of energy that appears literally out of nowhere, which appears to be linked to a force now propelling the expansion of the whole universe. And the evidence is more than just a few pretty pictures: it comes from decades of research in laboratories and observatories around the world.

The truth is that discoveries are now being made that prove beyond doubt that – just as the film claims – the universe is far stranger than anyone could have believed. Astronomers have found that the universe is made from an unknown form of matter, and is being propelled by a mysterious force known only as "dark energy". Meanwhile, physicists have discovered a bizarre phenomenon called "entanglement", in which atoms remain in intimate and instantaneous contact with each other, even if separated by billions of light-years. Many theoretical physicists now believe that our vast universe is just a tiny part of an infinite multiverse. Some even think the presence of parallel universes has already been detected in the laboratory.

The discoveries being made on a more human scale are no less astounding. Neuroscientists have found evidence that our conscious perception of events lags behind reality by around half a second – a delay we fail to notice because it is deliberately edited out by our brains. Anthropologists now believe they have identified the origins of modern humans, and how – and why – they left their birthplace to populate the world. And completing the cosmic circle, some theorists claim to have found links between the existence of humans – and indeed all life on Earth – and the fundamental design of the universe.

This book describes all these discoveries, and many more. Each of the twenty-five chapters is self-contained and they can be read in pretty much any order. As well as explaining the current state of play, each chapter also includes a glossary of technical terms, explanatory boxes and suggestions for further reading to allow readers to deepen, broaden and update their knowledge of fields that particularly interest them.

But my aim has been to do more than simply take readers up to the very frontiers of scientific knowledge. I have also sought to give some insight into the way science works, by describing the often tortuous route by which key discoveries have been made. All too often, the scientific process is portrayed as some kind of machine into which objective observations are poured, and out of which infallible truths emerge. As the chapters that follow

make clear, the reality is very different. Despite what some of its most distinguished practitioners might have us believe, science is a human endeavour, shot through with uncertainty and subjectivity – and is all the more fascinating for that.

Again, contrary to what some might have us believe, science shows no signs of reaching completion. On the contrary, we appear to be further away from omniscience than ever. It is now clear that many, if not most, natural phenomena can never be understood to the level once thought possible. The emergence of the concepts of chaos and quantum uncertainty have put ineluctable bounds on what we can know. Many of the chapters that follow outline techniques such as Bayesian inference and Extreme Value Theory which allow us to make the most of what we can know about the world.

Such techniques have application far beyond the quest for ultimate knowledge, however: Bayesian inference allows apparently impressive evidence of, say, some new cancer scare to be put in its proper context, while Extreme Value Theory underpins the design of the sea defences now protecting the 16 million people who live in The Netherlands.

One of the chapters in this book describes the efforts of some of the world's leading theoretical physicists to create the so-called Theory of Everything, which will sum up all the forces and particles of the universe in a single equation. My aim in this book is to confirm the suspicions of all who believe the universe is best summed up in a single word: magical.

Robert Matthews

OURSELVES – AND OTHERS

1
Consciousness

IN A NUTSHELL

Consciousness is something we all believe we possess, but saying exactly what it is has long challenged philosophers. Since the mid-1800s, scientists have found ways of probing brain activity, and linking it to traits we believe are vital for consciousness, such as free will and responding to stimuli. Studies of brain activity suggest that consciousness is just a very small part of brain activity, but one which is created from sensory input only after colossal effort. Experiments suggest that it takes around half a second for the brain to make us conscious of outside stimulation – though the delay appears to be "edited out" by the brain to keep it out of awareness. The result is a conscious mind with a model of reality that allows us to do more than merely react to stimuli or the bidding of our unconscious instincts – and turns us into sophisticated beings.

The room is full of people, yet totally silent. Sitting on cushions, their eyes are closed and their faces expressionless. They look totally absorbed in something – as well they might, for they are attempting to get to grips with one of the most profound mysteries in all science: the nature of consciousness.

As practitioners of Buddhist meditation, they are using techniques of mind-watching developed over 2500 years ago by the Indian philosopher Siddhartha Gautama, better known as the Buddha. The aim of these techniques is to turn the conscious mind in on itself and to watch it in action.

According to Buddhists, such introspection can give insights into the nature of mind, reality and the mystery of consciousness. Such claims have traditionally cut little ice with scientists, with their insistence on objective evidence. Yet now highly trained Buddhist monks are joining with scientists to probe the nature of consciousness. By summoning up mental states while undergoing brain scanning, the monks are opening up a new approach to what consciousness researchers call The Hard Problem: how does brain activity produce the experience of being conscious?

For something most of us are sure we possess, consciousness has proved amazingly hard to pin down. The seventeenth-century French philosopher René Descartes thought he had made a major advance by using a

TIMELINE

528 BC Indian philosopher Siddhartha Gautama makes study and control of consciousness the basis of a movement now known as Buddhism.

401 AD The philosopher and Catholic saint Augustine of Hippo identifies self-awareness as a key aspect of consciousness, declaring "I understand that I understand".

1637 French philosopher René Descartes puts forward his "dualistic" view of mind and body, arguing that the mind is not merely the actions of the brain.

1690 In his "Essay Concerning Human Understanding", the English philosopher John Locke defines consciousness as "the perception of what passes in a man's own mind".

1874 German psychologist Wilhelm Wundt moves consciousness out of purely philosophical inquiry, and advocates its study via introspection.

1890 Pioneering psychologist William James of Harvard University rejects Cartesian dualism, and concludes that consciousness is just a product of brain activity.

1913 American psychologist John B. Watson criticises attempts to study consciousness as hopelessly subjective, turning the field into a backwater for decades.

1929 Austrian psychiatrist Hans Berger invents electroencephalography (EEG), by showing that brain activity can be measured using electrodes placed on the skull.

1979 American brain scientist Benjamin Libet discovers the 0.5 second delay between brain activity and the conscious sense of deciding to act.

1988 Dutch-born psychologist Bernard Baars puts forward Global Workspace Theory, according to which consciousness is the process by which normally unconscious processes are brought together on a mental "stage".

1990s– present Advent of brain-scanning methods such as fMRI prompts huge increase of interest in consciousness by revealing brain activity in unprecedented detail.

logical argument to conclude that the conscious mind must be made of different stuff from brains and bodies, a distinction now known as Cartesian dualism. Yet even at the time, critics such as Baruch Spinoza pointed out that such a distinction raises profound problems about how mind and brain interact.

By 1690, the English philosopher John Locke had put forward the first working definition of consciousness as "the perception of what passes in a man's own mind". While this implied that consciousness was not so much a thing as an outcome of certain processes, what these were Locke could not say.

Not until the middle of the nineteenth century were scientists able to attempt an assault on the mystery of consciousness. The discovery of anaesthetics had revealed an intimate connection between body and mind – in flat contradiction of Descartes' claim. Researchers then set about tackling The Hard Problem, seeking ways of bridging the gulf between the subjective experiences of the mind and the objective study of brain activity.

In the 1860s, Wilhelm Wundt of Heidelberg University – now regarded as the father of experimental psychology – took the first tentative steps. Impressed by Spinoza's view that the conscious mind is a direct creation of bodily effects, Wundt set about trying to find out more about these effects. The technique he used was introspection, in which he trained students to note and describe their conscious response to outside stimuli.

Wundt's research highlighted the importance of understanding "qualia", the subjective experiences we have of the world around us – the "redness" of red, or the "sweetness" of sugar. Yet while Wundt worked hard to make his work objective, it was hard to gauge if one person's experience was the same every time, or matched anyone else's experience. He also lacked ways of making reliable objective measurements of brain activity that he could correlate with the subjective experience.

By the end of the nineteenth century, Wundt's research had convinced leading figures such as the influential American psychologist William James that consciousness was a direct outcome of brain activity, and thus worthy of scientific study. Yet many scientists sensed the existing techniques were simply not up to the task. Frustrated by the lack of hard results, most switched to more concrete problems, and the study of consciousness became an academic backwater.

It did not dry up completely, however, and over the next half-century scientists developed a variety of methods for tackling The Hard Problem. In 1929, the Austrian psychiatrist Hans Berger made the first breakthrough by finding a way of detecting electrical activity within the brain. Called the Electroencephalogram (EEG), it allowed Berger to discover two types of electrical activity, known as alpha and beta waves, that seemed to be linked with key aspects of consciousness. Alpha waves, oscillating around 10 times a second, appeared to reflect the state of consciousness, becoming fainter during sleep or anaesthesia. Beta waves, on the other hand, were about three times faster, and reflected concentration levels and non-conscious responses like the startle reflex.

Berger's discovery began the study of what are now called neural correlates of consciousness (NCCs): types of brain activity associated with conscious experience. These are now a

JARGON BUSTER

The Hard Problem: According to many, this is the central problem any theory of consciousness must solve: how to link the objective, physical structure of the brain to the subjective feeling of being conscious. The term was first coined in 1994 by the Australian philosopher David Chalmers at the University of Arizona.

Dualism: A view of the conscious mind as something fundamentally different from the living, working brain. Originally put forward by the seventeenth-century French philosopher René Descartes – and still called Cartesian dualism – the idea that the mind is more than mere brain activity is not widely accepted today.

Quale: The subjective experience of, say, redness or the softness of wool or the taste of lemons. Impossible to describe, yet regarded as essential to the concept of being conscious, some philosophers contend there is no such thing as a quale (or, in the plural, "qualia").

Global Workspace Theory: A proposal for how the brain works, according to which consciousness exists as a kind of mental "stage" – the workspace – where the inputs from other independent parts of the brain which work unconsciously are brought together. GWT was put forward in 1988 by the Dutch-born psychologist Bernard Baars of the Neurosciences Institute, San Diego.

Neural correlates: Actual parts of the brain and nervous system whose function can be directly linked to aspects of consciousness. Using brain-scans and laboratory experiments, some scientists claim that key features of consciousness lie in primitive parts of the brain such as the thalamus and brain stem.

Nerve impulses

The brain consists of around 100 billion neurons, each one receiving electrical impulses from many others via connections called dendrites, and transmitting its response to its neighbours via a single thread called an axon. The connections aren't seamless, however, and to bridge these gaps, nerve-endings are equipped with so-called synapses. These turn electrical signals into molecules called neurotransmitters that flow across the gap, triggering fresh electric impulses on the other side.

major focus of research by scientists, many of whom think that understanding consciousness involves understanding how the brain binds together a host of NCCs into a single, unified whole.

It is a belief spurred on by a surprising discovery made in the 1960s: that our consciousness involves only a tiny fraction of all the activity in our brains. A team led by the American neurologist Benjamin Libet applied very weak stimuli to the skin of patients whose brains had been exposed for neurosurgery. EEG measurements revealed that their brains had detected the stimuli – yet the patients themselves said they could feel nothing. It was the same story with stronger stimuli which lasted less than 0.5 seconds: while the brains of the patients detected it, the patients consciously felt nothing.

Similar findings have since emerged from studies of NCCs such as vision and the resulting qualia like the "redness" of red. Our eyes take in a torrent of information at the rate of around a megabyte per second, yet our consciousness seems to ignore all but a tiny percentage of it.

This huge disparity suggests that the brain performs a huge amount of unconscious processing of sensory input, distilling it down before we become conscious of it. Such

processing must take time to perform – suggesting there must be a time delay between our brains detecting a stimulus, and our mind consciously registering it. Attempts to measure this delay have led to perhaps the most startling discoveries yet made into the nature of consciousness.

In 1976, a team of researchers led by the German neurologist Hans Kornhuber set up an experiment to measure the time delay between the brain activity required to move a finger and actually making the movement. The speed of nerve impulses suggested the time delay would be around 200 milliseconds, similar to that of reflex actions. Yet the researchers found the delay was much longer. This was at least consistent with the idea that anything involving the conscious mind involves a lot of processing. The researchers found something else, however: the brain activity began around 800 ms before people finally got around to moving their finger. This was a startling discovery, with disturbing implications for the long-cherished notion of free will. For the sheer size of the delay hinted that our actions are not initiated by our conscious mind at all, but by the non-conscious brain activity taking place out of our perception.

An even more perplexing discovery was made in 1979 by Libet and his colleagues, during studies of the effect of applying direct stimuli to the brain.[1] Again, common sense suggested just a short delay between applying the stimulus and conscious detection – but again the researchers found a substantial delay, of around 500 ms. They also found something else: that the brain appears to "back-date" the conscious response, thus creating the impression there was hardly any delay at all.

These two findings not only cast new light on the link between brain activity and consciousness, but also gave some hints about what consciousness is actually for. First, although our actions are not initiated by our conscious mind at all, our consciousness can at least veto any actions generated by our non-conscious brain that we deem unacceptable. Free will is thus not about consciously choosing to act in certain ways, but about consciously choosing *not* to act.

Second, Libet's experiments point towards a reason why the brain expends so much effort to create consciousness: it binds together sensory inputs from the outside world to produce a consistent and reliable model of what is happening "out there".

This notion of consciousness as a model of reality fits in well with the sense we have of our brains creating a kind of mental "theatre". In 1988, the Dutch-born psychologist Bernard Baars took this idea to create the so-called Global Workspace Theory of consciousness. According to this, conscious processes are those currently in the "spotlight" of mental attention, while others remain out of the spotlight, stored in the memory for immediate access. Meanwhile unconscious processes are at work behind the scenes – and also form the mental audience, responding to what is currently in the spotlight.

GWT appears to be more than just a metaphor: it is based on results now emerging from the biggest breakthrough yet in the objective study of conscious processes: brain-scanning. Techniques such as functional Magnetic Resonance Imaging (fMRI)

Electroencephalogram

The invention of the Electroencephalogram (EEG) by the Austrian psychiatrist Hans Berger in 1929 was a major breakthrough in consciousness research, as it gave scientists a painless, non-invasive means to study the brain in action. Conscious and unconscious brain activity is the result of electrical signals flowing between individual brain cells, or neurons. While the signals between individual neurons are very weak, Berger found that the activity within specific parts of the brain is relatively easy to detect using pairs of electrodes placed around the skull. An EEG detects the signals as voltage differences between the pairs of electrodes, amplifies it and passes the result to a recorder, which captures the ebb and flow of brain activity.

give researchers detailed, real-time maps of brain activity, allowing it to be related to conscious processes. This has led to an explosion in studies of NCCs, with specific parts of the brain being identified as key players in conscious processes. For example, a central region known as the thalamus appears to be crucial in bringing sensory input into the "spotlight" of conscious attention, while the so-called ventro-medial cortex near the front of our brains seems to create our sense of life having purpose.

At the same time, researchers are beginning to look again at Wundt's methods for tackling the notoriously difficult subjective aspect of consciousness. They are recruiting people with decades of experience of examining their conscious states and reporting their experience: Buddhist monks. Early results from studies of monks undergoing brain-scans suggest that their years of intensive meditation allows them to produce stable mental states to order, giving researchers the consistency needed for reliable insights into the subjective experience of consciousness.

This meeting of cutting-edge technology and ancient spiritual practice may lead to new insights into the role of NCCs – and our ability to control them. Yet it fails to address some major mysteries about consciousness. Why do we possess it?[2] What advantages does it confer – and are humans alone in being fully conscious?[3]

One possible explanation lies in the view of consciousness as a means of creating a mental model of reality. Any organism possessing such a model can do more than merely react to stimuli, and pray the response is fast enough to escape predators. It can use the model to foresee threats and opportunities out there in the "real world" – thus freeing it of the speed limitations of non-conscious reflexes. A conscious creature, in other words, does not have to stumble around blindly, hoping its reflexes will keep it out of trouble. By binding together non-conscious responses to create even a simple model of reality, a creature possessing some degree of consciousness can avoid getting into tight corners in the first place – giving it a huge evolutionary advantage.

This in turn suggests that asking whether an organism is conscious or not may be the wrong question. Rather, consciousness may be a question of degree – with, say, an insect having a markedly less sophisticated model of reality than a human.

As with so many aspects of consciousness, definitive answers are still some way off. Even so, there is growing excitement that scientists are now closing in on the mystery of how 1,400 grams of squidgy tissue can endow us with our ineffable but unique sense of self.

Notes

1. In experiments performed on patients undergoing open-brain surgery, Benjmin Libet and his colleagues found that unless the brain is stimulated strongly enough for at least 500 ms, it fails to notice anything. In further tests, they stimulated the patients' brains, but this time followed it 250 ms later with a stimulus applied to their skin. As the 500 ms delay should apply in both cases, the patients were expected to report feeling the brain stimulus first, followed by the skin stimulus 250 ms later. Bizarrely, however, the patients reported feeling the skin stimulus first. Libet and his colleagues concluded that the brain "edits out" the 500 ms delay from the skin response to ensure that the actual time of the stimulus and the perception of it remain in sync. That, in turn, ensures that our mental model of reality – as constructed from our senses – isn't permanently 500 ms behind the times.

2. One of the most controversial theories of consciousness was put forward in 1989 by Professor Roger Penrose, an Oxford academic whose expertise lies in a field apparently unrelated to the mind: quantum mechanics, the laws of the sub-atomic world.

In a best-selling book *The Emperor's New Mind*, Penrose argued that the ability of human minds to make intuitive leaps towards deep truths about the universe is the result of a special type of computation triggered by quantum effects in the brain. These effects, he suggested, might also be responsible for binding together the activity in the brain to create a sense of a single, coherent consciousness.

In collaboration with the American anaesthetist Dr Stuart Hamerhoff, Penrose later argued that these quantum effects took place within tiny protein cylinders in nerve cells known as microtubules.

Despite attracting huge public interest, Penrose's theory has few adherents among consciousness researchers, who argue that it swaps one mystery for many more, and flies in the face of experimental evidence. Microtubules seem incapable of supporting the necessary quantum effects, and damage to them appears to have no effect on consciousness.

3. According to a survey published in 2001, around 1 in 10 people in Britain has experienced a so-called "out of body" experience (OBE), in which they felt as if their conscious being had separated from their body.

Many scientists dismiss OBEs as some bizarre glitch in brain function that creates the illusion of separation. If true, however, OBEs would present a profound challenge to current ideas of consciousness, which are inextricably linked to the physical brain.

To test the reality of OBEs, scientists have set up experiments to find out if people undergoing OBEs can see objects or numbers out of view of their physical bodies. Some of these have produced suggestive results, but nothing that has so far convinced mainstream scientists.

Further reading

Consciousness: an introduction by Susan Blackmore (Hodder & Stoughton, 2003)

The User Illusion by Torr Norretranders (Penguin, 1998)

Mapping the Mind by Rita Carter (Weidenfeld & Nicolson, 1998)

Understanding Consciousness by Max Velmans (Routledge, 2000)

How the Mind Works by Steven Pinker (Penguin, 1998)

The Astonishing Hypothesis by Francis Crick (Simon & Schuster, 1994)

2
Small World Theory

IN A NUTSHELL

People at parties often discover that they have surprising links with one another: a friend in common, say, or a colleague for whom they have both worked. Many of these "coincidences" are just that, or the result of people with similar backgrounds tending to move in similar circles. Yet there remains a surprising number of completely astonishing links that lead people to just shake their heads and reflect that "It's a small world". Until recently, little attempt was made to explain why a world of 6 billion people throws up such links so often. Initial studies in the 1950s hinted that random links played a key role, by "short-circuiting" the otherwise huge network of local communities that make up society. The breakthrough came in 1998, when mathematicians Duncan Watts and Steve Strogatz at Cornell University used computer simulations to show that just a tiny number of random, long-range links are enough to turn even a huge network of people into a "small world". Now researchers are finding evidence for an astonishing variety of small worlds, with implications for everything from the spread of disease to globalisation.

It has happened to all of us at some time or other: you're talking with a total stranger at a party, and you discover you have a friend or colleague in common.

Most people respond with a smile, exclaim "It's a small world!" – and think no more about it. The thing is, of course, the world isn't small at all: with 6 billion people, it's huge. And while most of us spend most of our time in our own little cliques, the "small world effect" pops up with surprising frequency.

For years, scientists tended to think it was all just coincidence. No longer: the effect is now one of the hottest topics in science, with researchers in fields as diverse as physics and economics taking it very seriously indeed. Understanding its cause and its consequences has led to the creation of a new field of science – Small World Theory – whose origins belie its major implications for issues ranging from the spread of disease to globalisation.

Its emergence highlights how a Big Idea that cuts across disciplines can go

TIMELINE

1951 Mathematician Anatole Rapoport and colleagues at the University of Chicago publish first studies of social networks and their effects.

1957 Ithiel de Sola Pool and Manfred Kochen in the US begin work on chains of acquaintances.

1959 Hungarian mathematicians Paul Erdös and Alfred Renyi prove the effectiveness of a few random links in connecting up huge networks.

1967 Stanley Milgram carries out first Small World experiment, posting letters to find out how many re-postings are needed to reach their destinations.

1990 John Guare's play *Six Degrees of Separation*, in which one of the characters points out that everyone is just six acquaintances away from everyone else, opens in America.

1993 Movie version of *Six Degrees of Separation*, starring Donald Sutherland.

1996 Research on chirping crickets lead Cornell University student Duncan Watts and thesis supervisor Steve Strogatz to study Small World Theory.

1996 Students at University of Virginia create the Web-based Kevin Bacon Game

1998 Watts and Strogatz publish a paper in *Nature* which sparks huge interest in Small World Theory.

1999 Watts publishes *Small Worlds*, collecting together key theory for the first time.

2001 Watts and colleagues launch Web-based version of Milgram's 1967 experiment.

2001 Damian Zanette uses Small World Theory to study disease spread.

While the "small world effect" has probably been noticed at gatherings for centuries, it was only in the 1950s that researchers began to probe its roots. At the University of Chicago, mathematician turned social scientist Anatole Rapoport and colleagues began by thinking of society as a "network" of individual people, each with random links to others. Some links were short, connecting up people into close-knit "communities", while others were relatively long.

Rapoport and his colleagues found that the random nature of the ties made a big difference to the structure of their artificial society. If the ties were made just a little less random, the society tended to fragment into isolated communities, with no links with people elsewhere.

It was an early hint of the crucial importance of random links in turning a huge population into a "small world", where everyone can be linked to everyone else via just a few links. Imagine, for example, that a society consists of a million people, each of whom has links only with ten people living nearby. If a rumour breaks out, it would take tens of thousands of steps for it to spread throughout the society, as it plods round from one "clique" of ten people to another. Now imagine the same society, where again everyone knows ten people – but this time randomly spread across the society. The rumour spreads far more quickly, popping up at random anywhere. After each re-telling, the number of people knowing the rumour grows by a factor of 10: first to 100, then 1000 and so on. After just six re-tellings, everyone will have heard the rumour – thanks to the random links that leap across the network.

unrecognised for years. Small World Theory may have its most striking applications in the "soft" social sciences, but its origins emerge from sophisticated mathematics and computer science. Only now, as the barriers between academic disciplines are breaking down, has Small World Theory been able to thrive.

Which is all very impressive, except that the real world isn't like that. We have ties that are neither totally random nor completely parochial, but a mixture of both. So why do we so often discover that we too live in a "small world"? In 1959, two Hungarian mathematicians, Paul Erdös and Alfred Renyi, took a big step towards answering the question. They came up with a formula showing how just a few random links can make a big difference. For example, in a crowd of 100 people you can form ties between virtually everyone by randomly picking just a dozen or so and introducing them to each other. Doing it systematically, on the other hand, demands that you arrange a staggering 4950 introductions: a few random links do the job far more effectively.

While Erdös and Renyi's formula showed the power of random links to connect people, it could not predict how "small" the real world might be. An ingenious experiment revealed in 1967 that the real world is astonishingly small.

Stanley Milgram,[1] a young social psychologist professor at Harvard University, had read of the work of two researchers in the US, Ithiel de Sola Pool and Manfred Kochen, which suggested that two strangers could be connected by short chains of acquaintances. Milgram decided to try to gauge the typical size of our social networks – how many people we have as friends, or friends of friends, and so on. To find out, he posted packages to 296 people in Nebraska and Boston, asking them to post them on to a "target" person in Massachusetts. Which sounds simple enough, except the recipients weren't told where the target person lived: only his name, occupation and a few other personal details. Milgram asked the recipients to post the packet to anyone they knew on first-name terms who might have a better chance of being able to deliver the packet.

The outcome was stunning: the packets typically reached the target after just five re-postings. A few years later, Milgram repeated his experiment, with

JARGON BUSTER

Regular network: A collection of objects or people who are linked to each other in a completely regular way – for example, only knowing their next-door neighbour. It typically takes many steps to link a person in one "neighbourhood" to anyone else.

Random network: A collection of objects or people who are linked to one another completely

randomly – for example, the star signs of people listed alphabetically in a telephone directory. The random connections usually result in relatively few steps being needed to link any one person to anyone else.

Small world: The type of network whose connections are partly regular and partly random. The result is typically lots of tight-knit cliques (due to the regular links), which

can still be reached by relatively few steps, thanks to the random links.

Characteristic path-length: The average number of jumps needed to get from one part of a network to another. For example, the "small world" of movies has a CPL of four, meaning any actor can typically be linked to any other via an average of just three other actors.

Kevin Bacon Game: A test of the knowledge of movie buffs, in which a contestant tries to link a named actor to Hollywood star Kevin Bacon via the fewest number of intermediaries. For example, Arnie Schwarzenegger can be linked in one step, via Colleen Camp, who appeared in *Last Action Hero* and with Bacon in *Trapped*.

Small worlds, big trouble

On 13 August 2003, a sagging power line touched a tree just outside Cleveland, Ohio, and triggered a dramatic demonstration of the small world effect. The resulting flash over tripped circuit-breakers that left 50 million people without power across eight US states, plus much of the eastern Canadian province of Ontario. The loss of power highlighted the existence of other "small worlds" linked to the electrical grid, including Canada's air and traffic networks, which were plunged into chaos. Then came the economic consequences: Canada's entire GNP dipped by 0.7 per cent, and US businesses lost an estimated $6 billion.

similar results: it seemed that everyone in America could be reached via just five re-postings. The implications were even more amazing. If just five postings were enough to reach anyone in a country of over 200 million, it seemed people typically knew around fifty others well enough to post the letter on. And that meant that just one more posting would be enough to reach anyone on the Earth. The real world, it seemed, really is a small world.

Milgram's findings have recently come under suspicion, with some researchers arguing that they aren't nearly as convincing as they seem. In 2002 psychologist Professor Judith Kleinfeld at the University of Alaska uncovered evidence that Milgram hand-picked data to back his claim. In both the original study and attempts to repeat it, most of the packages simply never arrived – raising doubts about the reliability of conclusions based on the handful that do. However, in 2003 researchers at Columbia University published results from a huge Internet-based version of Milgram's original experiment, and found that on average around six steps were needed to reach a target – backing Milgram's original claim.

Milgram's key finding has given its name to a play: *Six Degrees of Separation*, by the American playwright John Guare, in which one of the characters declares: "Everybody on this planet is separated by only six other people. Six degrees of separation. It's a profound thought … How every person is a new door, opening up into other worlds."

It was a profound thought that lodged in the mind of Duncan Watts, a 25-year-old PhD student at Cornell University in New York. In 1995, he was studying the mathematics of chirping crickets, and had run into a problem: how do the chirping crickets fall into step so quickly? Was each one listening to all his fellow crickets, or just to his closest neighbours? Then Watts recalled what his father had said years before, about how everyone in the world is just six handshakes away from the President of the United States. Watts wondered if the same phenomenon could be linked to the speed with which the crickets began chirping in unison.

Using an urban legend to solve a serious biological mystery is unconventional, to say the least, and Watts was nervous about suggesting the idea to his thesis advisor, mathematician Steve Strogatz. Yet far from laughing him out of his office, Strogatz thought the idea was intriguing, and the two began to collaborate. In June 1998, they published a paper summing up their findings in the prestigious journal *Nature* – and in the process created a whole new science: Small World Theory.

For the first time, Watts and Strogatz demonstrated the long-suspected power of a few random links to turn a sprawling world into a small one. Their success owed much to the ubiquity of computers. Unlike the pioneering theorists of the 1950s, Watts and

Strogatz could call on computer power to create simulations of the real world, with its mixture of close-knit communities and random links.

They created an artificial "society" of 1000 points, each one of which was connected to a clique of ten "friends". Watts and Strogatz found that if each clique consisted solely of nearest neighbours, then hundreds of steps were typically needed to get from one point to another. But when they broke just 1 in 100 of the close links and made them random, the number of steps needed plummeted ten-fold – while leaving most of the cliques well-connected.

Watts and Strogatz had finally revealed how a planet of 6 billion people so often seems a small world. While most of our friends may belong to our own little community, chance meetings have given us some randomly spread far more widely. And it is these handful of random links that "short circuit" the vastness of the global society and turn it into a small world.

Watts and Strogatz revealed the same effect at work in other "societies". For example, using a computerised database of actors and their films, they showed that Hollywood is another "small world". On average, any actor can be connected to any other via films involving just four intermediaries. Again, the reason is because of the links created by versatile actors like Rod Steiger, who worked in a huge if random range of films. Movie buffs had unwittingly exploited this for years when playing the so-called Kevin Bacon Game, where an actor has to be linked to another via the fewest number of films. Connect an actor to Rod Steiger, and things get much easier.

The paper by Watts and Strogatz triggered huge interest in finding other

Stopping the next pandemic

Small World Theory is likely to play a key role in the way the world tackles the next disease pandemic. According to the theory, most effort should be directed towards changing the "architecture" of links between infected and healthy people. This involves focusing resources on the relatively small number of highly "connected" people who are likely to spread infection most widely.

The importance of such people to the spread of disease was highlighted during the early days of the Aids epidemic. Investigators found that at least 40 of the first 248 men diagnosed with the disease were linked via sexual activity with one gay Canadian flight attendant, whose choice of partners and profession made him an extremely effective link in the disease network.

real-life examples of the effect. From the nervous system of worms to the network of computers making up the Web, small worlds have now been found in an astonishing variety of guises.

The implications are often startling. For example, the small world nature of the Web keeps it working despite having around 3 per cent of its crucial "router" computers down at any instant. Some other fast route can usually be found between any two computers, thanks to the relatively small number of random links that make the Web a small world.

But there's a flip-side to this: if one of these crucial random links becomes faulty, it can have a drastic effect on the whole Web. While the chances of it happening by fluke are small, such links present an obvious target for hackers or terrorists.

A similar effect has also been found in the business world. Major corporations have been shown to form a "small world", with apparently unrelated companies being connected by relatively few links. This can help the business world ride out the storms that always affect a proportion of companies – but it can also spell trouble if a key corporation with more or less random links elsewhere runs into trouble.[2]

Most disturbing of all are the implications of Small World Theory for the spread of diseases. Scientists have usually focused on the infectiousness of the disease to predict how epidemics spread. The small world effect shows that the society in which it breaks out can make all the difference. Studies by the Argentinian mathematician Damian Zanette suggests that if just 20 per cent of the population have random links to people beyond the site of the first outbreak, a minor outbreak can turn into a major epidemic.

Not surprisingly, some scientists now see this as crucial in the emergence of Aids in Africa. After the collapse of colonial rule, wars and the search for jobs triggered huge population movements – increasing the formation of random long-range links, turning what was once a localised viral disease into today's pandemic.

The importance of the small world effect for everything from the collapse of stock markets to the spread of disease is still the subject of much research. But perhaps the most important conclusion is already clear: that an apparently trivial bit of urban folklore can have implications that are anything but trivial.

Notes

1. Stanley Milgram (1933–1984) was one of the most brilliant and controversial scientists of the twentieth century. Even before his Small World research, he had achieved notoriety for his work on obedience to authority. In 1961, while still a junior professor at Yale University, he placed an advert in the *New Haven Register* in Connecticut, inviting readers to take part in a scientific study of memory. Those who took part were told it would focus on the effect of punishment on learning, and were led to a room to witness a man being wired up to electrodes that could, they were assured, deliver painful shocks. The recruits were then told to read out a list of word associations, and to give the pupil an electric shock whenever he made mistakes, using a console with switches going from 15 to 450 volts, marked "XXX".

   Although separated by a wall, the recruits could hear the pupil next door – and his cries of pain as he received shocks after each mistake. As his agony intensified, many of the recruits protested – only to be told by the scientist in charge that they must continue. And 65 per cent of them did – all the way up to "XXX", by which point the screams had been replaced by an ominous silence.

   Only once the experiment was over were the recruits told the truth: that the pupil was just an actor, and hadn't been harmed at all. Milgram had shown that ordinary people, from housewives to engineers, could be persuaded to abuse a perfect stranger to the point of death – if they believed they could pass responsibility to those in authority. In the 1960s, Milgram's experiment was seen as casting chilling light on the actions of the Nazis. As the recent scandal over the treatment of Iraqi prisoners shows, Milgram's experiment has lost none of its relevance.

2. Small World Theory casts intriguing light on the vexed issue of globalisation. In 1999, Bruce Kogut of the Wharton School of Business, and

Gordon Walker at the Southern Methodist University, Texas, used the methods devised by Watts and Strogatz to analyse the ownership networks of around five hundred of Germany's biggest corporations. Predictably, they found lots of "cliques" in the ownership of firms – the result of various tie-ups and mergers. But they also found that the short-circuiting effect of a few corporations typically allowed the ownership of any one firm to be linked to any other via just four intermediaries.

In other words, for all their apparent diversity, Germany's biggest corporations actually form a cosy small world. According to Kogut and Walker, this may explain why firms with apparently tenuous links to one another can still show similar corporate behaviour. But it may also have implications for the way these companies deal with globalisation. Small World Theory shows it can take only a few random links to "short-circuit" a vast network. So it's very likely that the whole corporate world has already become a small world.

In which case, Kogut and Walker conclude, there's little point fretting about globalisation – it may well already have had as big an impact on the business world as it ever will.

But there's a dark side to the corporate small world as well. If any part of it catches a cold, watch out – the "short circuit" effect could lead to collapses in apparently utterly unrelated businesses with astonishing speed.

The study of small worlds is still in its infancy, yet it's already clear that their presence in the real world holds both benefits and threats for all of us. Economists and business studies experts are likely to reveal many more examples of small worlds and their implications for us in the years ahead.

And some of what they find is likely to be disturbing. For while it may be fun to discover friends among perfect strangers, such encounters also serve as a warning – that the world is a more connected place than perhaps is good for us.

Further reading

Six Degrees: The Science of a Connected Age by Duncan Watts (Heinemann, 2003)

The Man Who Shocked the World: The Life and Legacy of Stanley Milgram by Thomas Blass (Basic Books, 2004)

"Collective dynamics of 'small world' networks" by Steven Strogatz and Duncan Watts in *Nature* vol 393 (1998) pp 440–2

Small Worlds by Duncan Watts (Princeton University Press, 1999)

3
Game Theory

IN A NUTSHELL

Game Theory started out as an attempt to find the best ways of playing games such as poker, in which players can't be certain of the intentions of their opponents. During the 1920s, mathematicians tried to solve the problem using the rule of choosing the strategy that gives the best payoff in the worst circumstances. In 1928, the Hungarian mathematician John Von Neumann proved that every two-player game has such a "minimax" strategy, as long as one player's gain exactly matched the other's loss. Many games like chess and poker have this "zero-sum" property; however, many real-life choices do not. For example, nuclear war is a non-zero-sum "game", as it can lead to neither side winning. Von Neumann's result was extended to cover such situations by the American mathematician John Nash, whose work showed that finding the best strategy often involves extra factors not covered by standard Game Theory – such as emotions. Attempts are now being made to extend Game Theory to include such factors, with implications for a host of fields, from world politics to crime prevention.

In October 1962 the world stood on the brink of nuclear Armageddon. The Soviet Union was setting up nuclear missiles in Cuba, just 145 km from the continental United States, and the Americans were demanding their immediate removal – or else. The missiles posed a threat which President Kennedy could not ignore, but he also knew that the wrong response could lead to nuclear war, and the death of millions. Some of his aides called for a massive airstrike, to take out the dozens of missiles on Cuba already pointing at the US. Yet this risked provoking the Soviet Union into launching an attack rather than risk losing its nuclear warheads. Others wanted a naval blockade, to prevent more missiles reaching the island, and demands for withdrawal – but some feared this would prove ineffective.

For a few fateful days, the two super-powers wrestled with their dilemma, all too aware of the consequences of making the wrong decision. President

TIMELINE

1713   English aristocrat James Waldegrave finds method for optimal play at the card game "le Her" which contains the basic Game Theory concept of a "minimax" strategy.

1921   French mathematician Emile Borel publishes papers on playing games where winning requires finding the best strategy when an opponent's thinking isn't known.

1928   John Von Neumann of Hungary proves the "Minimax Theorem", a key result of Game Theory for two-person zero-sum games.

1944   Von Neumann and the Austrian economist Oscar Morgenstern publish *Theory of Games and Economic Behaviour*, proposing a Game Theory approach to economics.

1950   Princeton student John Nash extends Von Neumann's original Minimax Theorem to cover non-zero-sum games, and coins concept of Nash Equilibria.

1955   Hollywood film *Rebel Without a Cause*, starring James Dean, includes the classic game-theoretic problem of playing "Chicken".

1973   British evolution theorist John Maynard Smith uses Game Theory to study how competing species can reach a stable equilibrium of populations.

1991   Veteran Game theorist Nigel Howard and others begin developing Drama Theory, which brings the role of emotions into Game Theory.

1994   Political scientist Professor Steven Brams at New York University introduces the Theory of Moves, which analyses how games evolve as each player responds to the strategies used by others.

1994   John Nash and other pioneers of Game Theory share the Nobel Prize for Economics.

The Cuban Missile Crisis is one of the turning-points of history. But it is also a nightmarish example of a problem we all meet countless times in everyday life: how to make optimal decisions when we aren't certain what the other person is thinking. From employees trying to negotiate pay rises to card-players wondering whether to bluff or fold, such dilemmas crop up time and again.

Could there be some way of working out the best way to play, the optimal strategy that leads to the best possible outcome? That is the question some far-sighted mathematicians began to ponder around a century ago. And they found the solution – in the process creating a new discipline called Game Theory, which has since found applications in areas ranging from military planning to poker, economics to evolution.

Despite its impressive reach, the origins of Game Theory lie in nothing more profound than the activity that gives the field its name: game-playing. As early as 1713, the English aristocrat James Waldegrave found a method for winning at a card game which contains many of the elements of modern Game Theory. Yet he failed to spot the potential applications elsewhere, and Game Theory had to wait another 200 years before finally taking off.

It was another card-player, the French mathematician Emile Borel, who made the connection between games and more serious problems. In 1921 he published the first of a series of papers about playing games where winning requires finding the best strategy when an opponent's thinking isn't known. Borel came up with a neat way of dealing with this lack of knowledge: play in such a way as to

Kennedy opted for a naval blockade, while also preparing for a massive strike against Cuba. A few days later, after feverish behind-the-scenes negotiations, the Soviet Union withdrew the missiles, and the world breathed again.

minimise the loss, regardless of how the opponent plays. Using this strategy, Borel was able to come up with rules of thumb for playing simple games, such as "Paper, Scissors, Stone", showing what mix of the three to use to minimise losses. Unlike Waldegrave, he also recognised that the same idea could have more serious applications in, for example, military strategy. Yet he also warned of the dangers of pushing such ideas too far – not least because he believed there was no way to find the best strategy for complex "games" where players face many options.

But Borel was wrong. Whenever two opponents battle for supremacy, and one player's gain is the other player's loss, there is always a best possible strategy that they should use. This is the so-called "minimax" strategy, and its existence was proved by a brilliant 25-year-old Hungarian mathematician named John Von Neumann.

Using extremely sophisticated methods, Von Neumann showed that the best strategy to use in such games is to study all the available options, work out their worst possible outcome, and then choose the least bad one. If either opponent tries to do better, they risk suffering a higher loss – making this "minimax" strategy the most rational choice. The strategy also solves the problem of trying to second-guess opponents: assuming they always act rationally, they too will pick the minimax strategy.

Von Neumann's proof of the Minimax Theorem made him the father of Game Theory. But he saw it as just the beginning. In 1944, together with the Austrian economist Oscar Morgenstern, he published *Theory of Games and Economic Behaviour*, which sought to make the Minimax Theorem the basis of a new approach to economics, which frequently involves two or more opponents competing to get the best possible outcome for themselves.

Von Neumann and Morgenstern's book showed the great potential of applying Game Theory to more than

JARGON BUSTER

Strategy: A way of playing a particular game, for example, play, bluff or fold in poker.

Evolutionary Stable Strategy: In biology, a type of behaviour that cannot be beaten if adopted by the population. Once adopted, Darwinian "survival of the fittest" will promote it.

Payoff: The gain or loss resulting from choosing a particular strategy to play a game. Sometimes replaced by the order of preference, for example one being the most-favoured, down to four in a two-player, two-strategy game.

Payoff matrix: A table of values showing the payoffs associated with each game-playing strategy available to the player.

Zero-sum game: A game in which any one player's positive payoff comes at the expense of the rest of the players, so that the total sum of all the payoffs, negative and positive, always tots up to zero.

Minimax Theorem: A method for finding the best strategy in zero-sum games with two players. Assuming both players want to do as well as possible, the most

rational strategy to choose is that which gives the best payoff even in the worst circumstances.

Nash Equilibrium: A set of strategies guaranteeing that every player can only do worse by choosing some other strategy. Named after John Nash, he proved that every game has at least one such set of strategies.

Are criminals here to stay?

Game Theory suggests we may always have to put up with a small proportion of criminals in society. According to psychologist Dr Andrew Colman at the University of Leicester, professional criminals have a choice of strategy: they can either do what they like, or conform. At the same time, the rest of us can either put up with criminals, or adopt their practices. The resulting combination of strategies and payoffs is then like the classic game of "chicken". Criminals and the rest of us would do best if the other party "chickened out" and co-operated; on the other hand, the worst possible outcome would be if everyone behaved like criminals. Using game-theoretic methods, Dr Colman showed that this leads to a stable state in which criminals taken out of circulation are replaced by formerly law-abiding citizens attracted by the niche left empty by the incarcerated criminals. The result is a permanent – if small – criminal element in society.

mere pastimes, and by the early 1950s US military strategists were using it to make sense of Cold War strategy. But they quickly discovered its limitations. The most serious was the assumption that one player's gain is always exactly matched by another's loss. Although many simple games like "Noughts and Crosses" obeyed this "zero-sum" rule, many real-life situations did not. For example, in dilemmas like the Cuban Missile Crisis, an attack by one side could prompt a nuclear war – which could result in annihilation for both players.

Von Neumann's theorem was silent about these "non-zero-sum" games. Was there a way of finding the optimal strategy for playing them? Come to that, did such a strategy even exist? Once again, a brilliant young mathematician found the answer, and again, it was positive. In 1950, a 21-year-old Princeton student named John Nash managed to extend Von Neumann's original Minimax Theorem to cover non-zero-sum games as well. Nash showed that for any game between any number of players, there is always at

least one strategy which guarantees that the players can only do worse by choosing anything else.

Known today as Nash Equilibria, these strategies are at the heart of Game Theory. They are also the source of enormous controversy. One reason is that these strategies don't always coincide with the obvious best choice for the players. An all too relevant example was the nuclear arms race that was underway when Nash made his discovery. Both the US and the Soviet Union knew the best option was to disarm, but neither trusted the other – so they both ended up spending vast sums on arms they hoped never to use.

It also became clear that many everyday situations had two or more Nash Equilibria – and it was far from clear which one players should pick. Again, there was a vivid contemporary example, in the 1955 Hollywood film *Rebel Without A Cause*, starring James Dean. Dean plays Jim, who takes on school bully Buzz in a game in which they race their cars toward the edge of a cliff, with the first one to "chicken out" losing. Jim and Buzz thus face a choice of swerving or driving on, giving the game of "Chicken" four possible outcomes – none of which was obviously ideal.[1] Swerving first means losing, but if both drive on the result is disastrous. Clearly it would be better if both decided to swerve – both lose, but both also get to live another day. Yet puzzlingly, Nash's work showed this "obvious" choice isn't a Nash Equilibrium: either player can do better by deciding to drive on while the other swerves. Worse, the game turns out to have two Nash Equilibrium strategies: driving on while the other player swerves, and vice versa. But how could the other driver be compelled to stick to the strategy?

Game theorists wrestled with this question, only to find more problems. It seemed something extra was needed. For example, Jim could pretend to be drunk before setting off, thus giving the impression death held no fear for him, and fooling Buzz into swerving first.

In short, Nash's work revealed that the clarity Von Neumann's original result brought to Game Theory was illusory. Since then, Game theorists have divided into two broad camps: one focusing on using classic games like "Chicken" to capture the essence of a problem; others seeking to extend standard Game Theory, and make it more realistic. Both have achieved impressive successes in fields ranging from economics to sociology. For example, Andrew Colman and colleagues at the University of Leicester have modelled the behaviour of criminals in society as a game of "Chicken". Both criminals and society would prefer the other party to "chicken out" first; on the other hand, neither side would benefit if everyone behaved like criminals. Using Game Theory, Colman and his colleagues showed that this leads to an essentially stable proportion of criminals in society, who do well enough from their activities without provoking society into draconian crackdowns. Game Theory also predicts that reducing crime below this level demands measures tough enough to convince criminals they are better off going straight – a prediction borne out by the success of zero tolerance policies in New York and other big cities around the world.

Following pioneering work by the British evolution theorist John

---

Making games more dynamic

The Theory of Moves (ToM) has been developed by political scientist Steven Brams as an extension of Game Theory which reflects the dynamic aspects of conflicts. Instead of regarding each conflict as an essentially static situation, ToM allows the responses of opponents to be included, producing a more dynamic and arguably more realistic framework in which to analyse conflicts. Brams and his co-workers have used ToM to re-evaluate many real-life conflicts with results that often contradict the predictions of Game Theory. For example, when it became clear that Israel was not going to be defeated in the Yom Kippur War of 1973, the Soviet Union – which was backing Egypt and Syria – offered to broker a settlement. The pro-Israeli Nixon administration responded by putting all US military forces on global nuclear alert. Game Theory predicts that this should have turned the war into a confrontation between the superpowers, but ToM shows that it was more likely to lead to peace – which, mercifully, is what transpired.

Maynard Smith, biologists use similar Game-theoretic ideas to understand why animals adopt certain types of behaviour, such as aggression or co-operation. Instead of a Nash Equilibrium, biologists talk of an "evolutionary stable strategy": behaviour which allows the population to resist invasion by others behaving differently. Von Neumann's original Minimax Theorem is even used in computer chess machines, where it helps select the handful of best moves from the huge numbers of possibilities. The most intriguing developments in Game Theory focus on making it ever more realistic. A group led by the British Game theorist Dr Nigel Howard, who advised the US government in the famous arms limitation talks in the 1960s, is developing so-called Drama Theory, which brings the role of emotions into Game Theory. Players who find themselves trapped in one kind of game often transform it into another game through their emotional response; Drama Theory attempts to predict the likely outcomes.[2]

One of the biggest advances in Game Theory has been made by political scientist Professor Steven Brams at New York University. He has developed an extension of Game Theory called Theory of Moves, which shows how games evolve as each player responds to the strategies used by others. It seems to give more plausible insights than standard Game Theory into key world events, from the Cuban Missile Crisis to the signing of the Good Friday Agreement in Northern Ireland.

The power of Game Theory is increasingly being recognised far beyond its original confines, reflected in the recent award of the Nobel Prize to a number of its pioneers, including John Nash in 1994.

The behaviour of people faced with tough choices may not be as simple as Von Neumann originally hoped, but there is little doubt that Game Theory has proved invaluable in unravelling some of its mysteries.

Notes

1. In both Game Theory and the Theory of Moves, conflicts are often analysed via a so-called "payoff matrix", which states the various options facing the various opponents, and their respective payoffs. For example, for the game of "chicken" portrayed in the film *Rebel Without A Cause*, the payoff matrix is as follows:

		Buzz's options	
		Swerve	Drive on
Jimbo's	Swerve	(3, 3)	(2, 4)
options	Drive on	(4, 2)	(1, 1)

Here the payoffs are given in "ordinal" form – that is, as rankings, with 4 being the best and 1 being the worst. The matrix thus reflects that fact that swerving can be pretty good for both (3,3), but both driving on will be disastrous (1,1). The outcome of any game is then assessed by considering how the players will respond to the various combinations of payoffs – and, in the Theory of Moves, whether they will want to move to other options in the search for something better.

2. Criticism of Game Theory often centres on its assumption that "players" always

behave rationally. Nigel Howard has illustrated the dangers of the rationality assumption with the real-life tale of two economists taking a taxi to their hotel in Jerusalem. Worried that they were going to be overcharged, they decided not to haggle about the price until they'd reached the hotel, when their bargaining position would be much stronger. But their entirely rational, game-theoretic strategy didn't work out too well. Realising he'd been backed into a corner by these two smart alecks, the driver locked the taxi doors, drove them back to where they'd started – and dumped them on the street.

Further reading

Mathematics and Politics: Strategy, Voting, Power and Proof by Alan D. Taylor (Springer, 1995)

Theory of Moves by Steven J. Brams (Cambridge University Press, 1994)

Prisoner's Dilemma: John von Neumann, Game Theory and the Puzzle of the Bomb by William Poundstone (Oxford University Press, 1992)

DOING THE IMPOSSIBLE

4
Artificial Intelligence

IN A NUTSHELL

Since the 1950s, scientists have been developing ways of allowing computers to do more than simple number-crunching tasks. To imbue them with "artificial intelligence" (AI), computers have to be programmed in such a way that they can carry out human-like tasks such as reasoning from evidence, and recognising patterns. A range of AI techniques have now emerged to tackle such challenges, which use special programming languages that allow computers to acquire human-like expertise and apply it to new problems. These breakthroughs have produced some impressive successes, ranging from diagnosing diseases from symptoms to spotting wanted criminals on CC-TV images. Some specialised AI computers have even created genuine works of art, made their own scientific discoveries – and beaten human world champions at both chess and draughts. Even so, computers still have a long way to go before becoming truly "intelligent" and passing the so-called Turing Test, according to which computers can convince people that they are actually human. One of the biggest problems lies in getting computers simply to understand ordinary language, with all its ambiguities.

Like all renowned artists, Aaron has a unique style – the result of over 20 years spent mastering colour and composition. With bold strokes and vibrant colours, Aaron works quickly, creating images that will sell for thousands of pounds around the world.

Not that Aaron thinks about the money. In fact, it's debatable whether he thinks at all – for Aaron is a computer. Strictly speaking, Aaron isn't even that:

he – or rather it – is just thousands of lines of Lisp, a programming language that can imbue a computer with that most controversial of traits: Artificial Intelligence (AI).

Since the early 1970s, the British-born artist and computer scientist Harold Cohen of San Diego University, California has been giving Aaron skills that allow it to create unique works without any human assistance. Is

TIMELINE

1943	Warren McCulloch and Walter Pitts create electric "neural networks".
1950	Alan Turing creates his eponymous test for a thinking computer.
1951	Marvin Minsky creates an electronic "rat" that learns to escape from a maze.
1956	Dartmouth College Conference, where John McCarthy coins the term "Artificial Intelligence".
1956	Herbert Simon and colleagues introduce first AI program.
1960	Frank Rosenblatt at Cornell demonstrates the Perceptron.
1969	Minsky and Seymour Papert's criticisms kill off neural network research.
1972	Harold Cohen begins work on Aaron, the computer artist.
1976	Douglas Lenat of Stanford University creates AM, a program which makes mathematical discoveries.
1982	John Hopfield of Caltech revives neural net research.
1997	IBM computer Deep Blue beats world chess champion Garry Kasparov.
Late 1990s–present	AI methods start to become mainstream.

many regard as the father of AI. In 1950, a few years before his suicide at the age of 41, the brilliant Cambridge mathematician, code-breaker and computer pioneer Alan Turing had a vision of the future. He prophesied that by the end of the twentieth century computers would be able to hold a five-minute conversation with humans, and fool 30 per cent of them into believing they were dealing with another human being.

It is a deadline that has now come and gone, yet despite spending billions of pounds in funding, no computer has come close to fooling its creator that it is really "thinking". So have computer scientists failed in their quest ? Not at all: indeed, in some cases like Aaron, they have arguably succeeded in giving machines abilities that can pass for those of a human.

More importantly, most AI researchers have come to see Turing's vision as an irrelevance compared with a far more important goal: creating machines that can relieve humans of the drudgery of, say, scanning medical slides for cancerous cells or spotting fraud in financial records.

Even so, in its earliest days AI research was driven by the belief that human-like traits could soon be matched and even surpassed by machines. That optimism stemmed from a key discovery made by researchers 60 years ago: that even crude "electric brains" could perform impressive feats.

In 1943, two American neurologists named Warren McCulloch and Walter Pitts stunned fellow researchers by showing that it was possible to understand the action of nerves using the laws of mathematical logic, which underpin the basic processes of

Aaron thinking? Is he creative? These are questions central to the field of AI, and its quest to give machines the power of the human mind.

It is a quest whose fortunes have soared and sunk several times since it began over half a century ago. Now, at the start of the twenty-first century, AI is starting to work its way into everyday life. But it is doing so in ways far removed from the sci-fi imagery it once conjured up, of human-like robots and malcontent computers like HAL in *2001: A Space Odyssey*.

It has also taken half a century to shake off the ghost of the man who reasoning. They also showed that nerves could be mimicked using electric circuits wired up in certain ways – thus opening up the prospect of creating artificial brains that could reason and "think".

With real brains containing billions of nerve cells, it looked likely that anything mimicking their power using electric components was centuries away. Yet researchers soon discovered that even relatively simple "neural networks" could be trained to solve surprisingly difficult tasks.

The trick lay in wiring them up so that their response to particular inputs could be tuned to give the right output. By 1951, the American computer scientist Marvin Minsky had wired together hundreds of vacuum tubes so they could be taught to mimic a rat's ability to navigate a maze.

The way ahead now seemed clear: just build ever bigger neural networks, and train them up. Not everyone was convinced, however. The problem with such networks was – and is – that it's never clear exactly how or why they act as they do. Opening them up to examine their connections is all but useless.

Such concerns prompted some researchers to take a different route towards AI, based directly on the rules of logic. The idea was to create machines whose circuits would mimic the reasoning ability of the human mind. By the mid-1950s, Herbert Simon of the Carnegie Institute of Technology in Pittsburgh and colleagues had succeeded in capturing the laws of logic in a format that computers could understand.

The result was the Logic Theorist, whose abilities were no less impressive than those of Minsky's electric "rat". Despite running on a relatively crude computer, the program was able to prove dozens of theorems about the foundations of mathematics – a task

JARGON BUSTER

Neural computing: A technique for programming ordinary computers to behave like networks of very crude "brain cells", which can be trained to recognise patterns. Widely used in tasks such as identifying faces on security camera images, and in speech recognition.

Expert Systems: Computers programmed to capture the skill of a human expert and use that expertise to perform "intelligent" reasoning, using mathematical logic. The human expert's skills are stored in the so-called "knowledge base", which the computer exploits in reaching its conclusions. Used in tasks like diagnosing diseases from symptoms.

Turing Test: A controversial test of computer intelligence, based on ideas put forward in 1950 by the pioneering British computer scientist Alan Turing. According to this test, if you're using a computer and cannot tell the difference between its responses and those you would expect from a human, then you can regard the computer as "thinking".

Natural Language Methods: Techniques allowing humans to interact with computers as if they were fellow humans – using ordinary language rather than specialised programming languages. Requires the computer to cope with the ambiguities and context-dependence of ordinary conversation.

Predicate Logic: Extension of ordinary rules of logic, with their focus on "true/false" statements, to capture relationships between concepts and objects. Widely used to give computers the ability to reason about complex situations, e.g. why someone should be given a loan.

The Anthropomorphic Trap

Each year AI researchers compete for the Loebner Prize, set up in 1990 by an American philanthropist and offering $100,000 and a medal to the designers of the first computer to convince judges that it is passably human. While no computer has yet come close to claiming the prize, the sum of $2000 and a bronze medal is awarded each year to the computer program that does best, some of which are available online (see www.tinyurl.com/9men4). Interacting with them often highlights a trait noted since the earliest days of AI research: our penchant for anthropomorphism. In 1966, Joseph Weizenbaum of the Massachusetts Institute of Technology unveiled Eliza, a computer program that gave an eerily accurate impersonation of a psychotherapist. Its success came from tricks such as turning conversations back towards patients whenever possible – thus compelling them to do all the thinking. Weizenbaum found the tendency of people to pour out their hearts to it pretty disturbing: "Extremely short exposures to a relatively simple computer program", he noted, causes "powerful delusional thinking in quite normal people".

thought to be the preserve of brilliant human logicians.

The Logic Theorist is now regarded as the first Artificial Intelligence program – and made its debut in 1956 at the first-ever AI conference, held at Dartmouth College, New Hampshire. It was a conference suffused with excitement over the prospects then emerging – and saw the birth of the term "Artificial Intelligence". The following year, Herbert Simon was predicting the emergence within 10 years of computers so smart they would be able to make their own mathematical discoveries and become world chess champion.

By the early 1960s, neural networks were going from strength to strength. At the Cornell Aeronautical Laboratory in New York, psychologist Frank Rosenblatt had created the Perceptron, a network of electronics which had the uncanny ability to "remember" patterns of lights. Like a human putting a name to a familiar face, the Perceptron could be trained to recognise certain patterns of lights.

It was an impressive achievement, and one that prompted Rosenblatt to tout the Perceptron as the prototype of a truly "thinking" machine. Inevitably, however, such grand claims annoyed his rivals – not least because it threatened their chances of getting some of the huge funds being offered by the US Department of Defense for AI research.

In 1969 two of Rosenblatt's rivals, Seymour Papert and Marvin Minsky of the Massachusetts Institute of Technology, published a book claiming to prove that the future of AI could not lie with the Perceptron. They showed that it lacked the ability to solve a type of logical problem important in AI applications, at least in its simplest format. They also insisted that there was no way of modifying the Perceptron to allow it to succeed. This was soon shown to be false, but by then Papert and Minsky had succeeded in choking off funds not only to the Perceptron but all neural network research.

Until the mid-1980s, the route to AI was dominated by logic-based programs running on conventional computers. Attention focused on so-called expert system methods, in which computers were imbued with the skills of human experts in, say, medical diagnosis or fraud detection. This involved setting up a "knowledge base" of rules, based on structured interviews with the human expert, and programming computers to exploit these rules, applying logic to reach decisions about, say, how best to treat a cancer patient.

As so often in AI research, however, the reality fell far short of the hype. Capturing human expertise proved harder than many thought, while the computers of the day had problems coping with the demands of logical reasoning. Nor was everyone keen to welcome these computerised "experts" into their area. Reports of their failure were seized on with glee, while companies who found them useful kept quiet, lest rivals discover the secret of their success.

By the mid-1980s, it was clear that expert systems could never be the panacea some had claimed. But by then interest in neural networks was starting to re-emerge. John Hopfield, a theoretical biologist at Caltech, revived interest in the links between neural networks and living cells, while others pointed out ways around Papert and Minsky's supposedly knock-out argument against Perceptrons. With the in-fighting now more or less over and cheap computing power available in abundance, the AI field settled down to attacking real-world problems.

Their variety is breathtaking. The logical reasoning abilities of expert systems have found application in fields ranging from fraud detection in banking to picking out gene sequences from the billions of chemical letters of DNA. Neural networks, meanwhile, have found success in applications which play to their strengths: pattern-detection and learning from experience. They are now being used to keep watch for suspect cars approaching London, to spot wanted criminals in high-streets – and even to detect signs of potential violence in crowds before it breaks out.[1]

Yet for all their usefulness, today's real-life AI systems seem a far cry from the "thinking machines" once promised. Whatever happened to Alan Turing's vision of a computer that could convince humans it was one of them, or Herbert Simon's 1957 prediction of computers making math-

The computerised Shakespeare scholar

Like the human brain, neural networks are especially good at spotting patterns hidden among confusing detail. This has led to some unusual applications for this type of AI technology, including stylometry – the identification of authors from their writing style. In the early 1990s, the Shakespeare scholar Dr Tom Merriam and I developed a neural network that can recognise the writing style of Shakespeare and his contemporaries. It works by extracting stylometric features, such as the relative frequencies of certain words, from large samples of text and associating them with each author. Once trained, the neural network proved very adept at identifying the authors of texts it had not seen before. It also cast intriguing new light on the mystery of the Bard's transformation from jobbing actor to dramatist of genius. Some of his earliest plays bear signs of being amended versions of scripts written by his great contemporary Christopher Marlowe.

ematical discoveries, and becoming world chess champion?

In fact, they all came true – at least, up to a point. In 1966, Joseph Weizenbaum of MIT unveiled Eliza, a computer that could respond to humans using ordinary language. Weizenbaum hoped Eliza would break down the barriers between computers and humans. To his horror, he found humans were all too keen to break down those barriers themselves. Eliza was programmed to give responses like those expected from a psychotherapist, with statements like "I have problems with my father" triggering Eliza to respond with "Tell me more about him."

Weizenbaum discovered that people quickly opened their hearts to his machine. He was deeply worried by the fact that although Eliza clearly doesn't "think", all too many humans would allow her to pass the Turing Test.

Simon's prediction that a computer would make a mathematical discovery came to pass in 1976, when a logic-based program named AM, developed by Douglas Lenat at Stanford University, claimed that every even number greater than four is the sum of two odd primes. A human mathematician named Christian Goldbach had made the same claim over 200 years earlier; even so, the rediscovery of "Goldbach's Conjecture" by AM gave a glimpse of the potential of AI.[2] An even more impressive demonstration came in 1997 when IBM's Deep Blue computer beat Garry Kasparov, the greatest human chess player in history.

While spectacular, these successes are the result of one-off projects, using methods far from the mainstream of AI research. Certainly they cast little light on how the 1300 grams of squishy grey matter we call the human brain can do all these things, and far more besides.

Fifty years after it began, it is clear that the quest to capture that astonishing power in a computer remains as challenging as ever.

Notes

1. Most of us curse junk mail, but it would be even more irksome if not for AI methods, which have ensured that most of it is now pretty well-targeted. This has been made possible by the use of AI in so-called "data mining", in which computers are used to look for patterns in the vast quantities of data we generate simply by going about our daily lives. Using sources such as supermarket checkout data, AI techniques are able to deduce a profile of who we are, what we want and when – and target promotional literature accordingly.

2. The origins of AM's apparent creativity have been the source of considerable controversy. In 1984, British AI experts Keith Hanna and Graeme Ritchie published a study of AM which showed that it may have benefited from ideas unconsciously built into its code by Lenat to help it discover new ideas, and

also from Lenat identifying promising ideas as they emerged from AM. In addition, Lenat himself pointed out that Lisp, the computer language used to create AM, has its origins in the so-called lambda calculus – which was invented specifically to probe the foundations of mathematics. Lenat came to suspect that AM's mathematical successes were chiefly the result of the strong correlations between Lisp and mathematics.

Further reading

The Brain Makers by H. P. Newquist (Sams Publishing, 1994)

Artificial Intelligence: A Beginner's Guide by Blay Whitby (Oneworld, 2003)

Cognizers: Neural Networks and Machines that Think by R. Colin Johnson and Chappell Brown (Wiley, 1988)

5
Information Theory

IN A NUTSHELL

From DVDs and hard drives to satellite communications and text messaging, information plays a key role in virtually every aspect of our lives. In 1948, an American engineer named Claude Shannon laid the foundations of what is now called Information Theory, which showed the limits to rapid yet reliable communications. It led to the development of techniques such as data compression and error correction, which allow data to be stored and transmitted as efficiently as possible. But Information Theory is proving to have major application in far deeper problems as well. The genetic code of DNA turns out to be an elegant example of information storage, and researchers are using Information Theory to decode the genomes of living organisms. Meanwhile, physicists have found evidence that information lies at the heart of the long-sought Theory of Everything. In short, Information Theory encompasses life, the universe and everything.

Nothing like it had been attempted before, and even NASA's experts wondered if it could really work. But after monitoring the data, there was little doubt that they had to do something – or else lose contact with their space probe forever. Launched in 1977, Voyager 1 had sent back spectacular images of Jupiter and Saturn and then soared out of the solar system on a one-way mission to the stars. But after 25 years of exposure to the frigid temperatures of deep space, the probe was beginning to show its age. Sensors and circuits were on the brink of failing, and with the probe 12,500

billion kilometres from Earth there seemed nothing anyone could do. Unless, that is, the NASA engineers could somehow get a message to Voyager 1, instructing it to dust off some spares and use those instead.

In April 2002, one of the huge radio dishes belonging to NASA's Deep Space Network sent the message out into the depths of space. Even travelling at the speed of light, it took over 11 hours to reach its target, far beyond the orbit of Pluto. Yet, incredibly, the little probe managed to hear the faint call from its home planet, and successfully made the switch-over.

TIMELINE

1941 Claude Shannon joins Bell Laboratories in New Jersey and begins laying the foundations of Information Theory in a paper which was published in 1948.

1947 US statistician John Tukey invents the term "bit" for the smallest element of information needed to describe the state of a two-state system (like "on" or "off").

1966 Identification of combinations of DNA "bases" as genes for building proteins shows use of Information Theory in living organisms.

1969 United Nations agreement leads to introduction of International Standard Book Number, a 10-digit description of books based on Information Theory principles

1973 A fundamental link between information and black holes is uncovered by physicist Jacob Bekenstein, who shows there is a limit on their storage capacity.

1974 Wrigley's Gum becomes first product featuring a Universal Product Code ("barcode") invented the previous year by George Laurer of the US.

1992 Eugene Stanley of Boston University uses Information Theory to show "junk" DNA may hold non-genetic information vital to organisms like humans.

1997 Theorists Stephen Hawking and Kip Thorne set up a bet with John Preskill at Caltech that information is destroyed inside black holes.

1999 Physicist Roy Frieden proposes that all the laws of physics may ultimately be the result of exchange of information between the observer and the universe.

2002 Scientists successfully repair Voyager 1 spacecraft at a distance of 12,500 billion km using error-correcting signals.

2004 Team led by Samir Mathur at Ohio State University uses so-called superstring theory to predict that information survives inside black holes.

It was the longest distance repair job in history, and a triumph for the NASA engineers. But it also highlighted the astonishing power of techniques developed by an American communications engineer who had died just a year earlier, named Claude Shannon. In the 1940s, he had single-handedly created an entire science of communication which has since found its way into a host of applications, from DVDs to satellite communications to barcodes – anywhere, in short, where data has to be conveyed rapidly yet accurately.

Known as Information Theory, it underpins many of today's most important technologies. But now a whole new area of application is starting to emerge – one with profound implications for the very nature of space and time. Some of the world's leading physicists believe that Information Theory holds the key to understanding some of the most profound mysteries in the cosmos, from the nature of black holes to the very meaning of reality.

This all seems light-years away from the down-to-earth uses Shannon originally had for his work, which began when he was a 22-year-old graduate engineering student at the prestigious Massachusetts Institute of Technology in 1939. He set out with an apparently simple aim: to pin down the precise meaning of the concept of "information". The most basic form of information, Shannon argued, is whether something is true or false – which can be captured by a single binary unit or "bit", of the form 1 or 0. Having identified this fundamental unit, Shannon set about defining otherwise vague ideas about information and transmitting it from place to place. In the process he discovered something surprising: it is always possible to guarantee messages get through random interference – "noise" – intact. The trick, Shannon showed, is to find ways of packaging up – "coding" – information to cope with the ravages of noise, while still staying within the information-carrying capacity – "bandwidth" – of the communication system.

Over the years, scientists have devised many such coding methods, and they have proved crucial in many technological feats. The Voyager spacecraft transmitted data using codes which added one extra bit for every bit of information; the result was an error rate of just 1 bit in 10,000 – and stunningly clear pictures of the planets. Other codes have become part of everyday life – such as the Universal Product Code, or "barcode", which uses a simple error-detecting system that ensures supermarket check-out lasers can read the price even on, say, a crumpled bag of crisps. As recently as 1993, engineers at France Telecom made a major breakthrough by discovering so-called Turbo Codes, which

come very close to Shannon's ultimate limit and now play a key role in the mobile videophone revolution.

Shannon also laid the foundations for efficient ways of storing information, by stripping out unnecessary – "redundant" – bits from data which contributed little real information. As mobile phone text-messages like "I CN C U" show, it is often possible to strip out a lot of data without losing much meaning. As with error-correction, however, there's a limit beyond which messages become too ambiguous. Shannon showed how to calculate this limit, opening the way to the design of compression methods that cram maximum information into minimal space.

Not surprisingly, Shannon's publication in 1948 of A *Mathematical Theory of Communication* was quickly recognised as a turning-point in technological history. Yet Shannon himself refused to take part in what he saw as hype. Ironically, hints were already emerging that Information Theory was

JARGON BUSTER

Bit: A term first coined by the American statistician John Tukey, a "bit" is the smallest possible unit of information, which simply describes the state of something that can be in just one of two states – for example, on/off.

Bekenstein Bound: The absolute maximum amount of information – expressed as "bits" – that can be stored or transferred in any given

region of space. Rough calculations suggest the limit is about 10 to the power of 70 bits per square metre – by comparison, CDs cram only about 10 to the power of 13 bits per square metre.

Redundancy: Extra bits in communication that add little extra information. Although apparently useless, such redundancy is important in combating the effects

of noise. For example, although vowels are often "redundant" in English, including them helps prevent garbling of messages.

Data compression: technique used to cut out superfluous – "redundant" – bits from data to make it faster to transmit and more compact for storage. Investigated by Shannon in his seminal 1949 paper, the technique is

now widely used in everything from satellite communications to DVD players.

Information Capacity Theorem: Key result of Information Theory proved by Shannon, which shows that communication is always a compromise between sending messages rapidly, and ensuring they are error free.

Turbo Codes

In 1993, two French professors amazed communications experts by unveiling a way of transmitting signals at speeds and error-rates deemed impossible. Engineers had spent decades searching for ways of packaging up – "coding" – signals to get as close as possible to the ultimate limit of fast, accurate communication set by Shannon's Information Capacity Theorem, yet even the best codes typically demanded over twice the signal strength predicted by the theorem. Claude Berrou and Alain Glavieux of the Ecole Nationale Supérieure des Télécommunications de Bretagne in Brest astounded experts by finding codes that need just a few per cent more power than Shannon's ultimate limit. Known as Turbo Codes, they work by using two separate coders and decoders on the same signal, allowing each to check the work of the other.

even bigger than even its most enthusiastic advocates believed.

At the time, many scientists were struggling to solve the mystery of biological information – specifically, the precise nature of the instructions used by living organisms to create offspring. Researchers talked vaguely of "genes" but no-one knew what they really were. Many believed they had to be proteins, which appeared to have the complexity to act as the instructions for living organisms. But in 1944, the American biochemist Oswald Avery and colleagues surprised many by showing that genetic information was carried by a far simpler molecule in cells known as DNA.

It took another 20 years to work out how information is carried by the four chemical "bases" of DNA, but it turned out that genes had evolved according to the principles of Information Theory. For complex organisms like humans, however, genes make up only a few per cent of DNA, the rest being apparently useless strings of bases dubbed "junk" DNA. In 1992 Eugene Stanley of Boston University used a technique based on Information Theory to discover hints of "redundancy" within

junk DNA, suggesting there is information lurking in these non-genetic regions. Growing numbers of researchers now suspect that the success of genomic medicine will depend on understanding the impact of the information in junk DNA.

Information Theory is also emerging as a key part in the quest to find the holy grail of theoretical physics: the Theory of Everything (ToE). We will deal with the subject in detail in Chapter 21, but put simply, the ToE aims to sum up in a single set of equations the origin and nature of the cosmos, and the forces within it. The biggest obstacle has been combining the two most successful theories in science: Quantum Theory, which governs the sub-atomic world, and Einstein's General Theory of Relativity, which describes the nature of gravity. These two theories could hardly be more different. According to Quantum Theory, gravity is transmitted from place to place by "carrier particles", while Relativity sees gravity as the result of the curvature of space and time. Somehow a ToE must unify these two radically different conceptions – but how?

Hints that Information Theory may hold the key emerged in the early 1970s, following research into perhaps the most bizarre objects in the universe: black holes. Formed whenever objects such as supermassive stars collapse under their own gravity, black holes cram so much mass into so little space that their gravitational fields are incredibly intense – so intense, indeed, that not even light can escape their clutches.

In 1973, a young graduate student at Princeton University named Jacob Bekenstein began studying the implications of the prodigious appetite of black holes. Every time an object disappears inside them, information goes in with it, but Bekenstein proved there is a limit to the amount of information a given black hole can contain. The amount is unimaginably vast: even a black hole the size of a proton can contain as much information as trillions of CDs. But what intrigued scientists was the formula that gave this maximum limit, known as the Bekenstein Bound. Contrary to common sense, the formula showed that the limit does not depend on the volume of the black hole, but just on its surface area. Secondly, the answer it gives is expressed in terms of so-called Planck Areas. These are the smallest conceivable areas of space – and are expected to play a key role in any Theory of Everything.

Bekenstein's formula, in other words, seemed to hint at connections between the long-sought ToE and information. In the last few years, other theorists have found more such hints. Different approaches to the problem of unifying Quantum Theory with Einstein's concept of gravity have been found to lead to Bekenstein's formula – boosting confidence that information will prove important in the ultimate ToE.

In 1999, physicist Roy Frieden at the University of Arizona uncovered startling evidence that the very laws of physics are ultimately about extracting information from the universe. Theoretical physicists have long been intrigued by the fact that everything from Newton's laws of motion to the rules of quantum mechanics can be derived from the same recipe. Put simply, by inserting the right "key" into a mathematical machine known as the Principle of Least Action, it's possible to

The Principle of Least Action

First formulated on philosophical grounds by the French mathematician Pierre Maupertuis in 1744, the Principle of Least Action states that nature always acts so as to to minimise the amount of "action" involved, a quantity that depends on the properties of particles. Physicists later discovered that the correct choice of action allows entire laws of physics to be extracted from the principle, but why it works is still deeply controversial. Recent research suggests it may be linked to Information Theory.

derive all the laws of physics. But while theorists have been happy to use it, they have been unable to explain why or how this machine works its magic.

Frieden showed that the answer may lie in information. Whenever we try to understand how the cosmos works, we are asking a question: where will this particle go, for example, or how fast will it move? Frieden used Information Theory to find a way of obtaining the "most informative" answers to such questions – and it turns out those answers are what we call the laws of physics.

Although Information Theory is casting light on many profound questions, many mysteries remain – such as what happens to information swallowed up by a black hole. It seems obvious that any such information must vanish for good, yet this violates fundamental principles underpinning Quantum Theory. In 1997 this so-called Black Hole Information Paradox led Stephen Hawking of Cambridge University to make a bet with some colleagues that Quantum Theory is flawed, and information really does vanish inside black holes. Hawking has since decided to pay up, having found what he at least believes are compelling arguments showing information does survive inside a black hole after all.[2]

No one knows for sure; indeed, it might never be possible to find out. The real importance of such arguments lies in the hints they give about the role of information in creating the ultimate Theory of Everything. And the evidence so far leads to an astonishing

possibility: that the keys to the cosmos may be intimately linked to the same principles used to design the bar-codes on your groceries.

Notes

1. Noise usually means unwanted sounds which interfere with genuine information. Information Theory generalises this idea, via theorems that capture the effects of noise with mathematical precision. In particular, Shannon showed that noise sets a limit on the maximum rate at which information can pass along communications channels while remaining error-free. This rate depends on the relative strengths of the signal and noise (the "signal-to-noise" ratio) travelling down the communication channel, and on its information-carrying capacity (its "bandwidth"). The resulting limit, given in units of bits per second, is the absolute maximum rate of error-free communication possible for given signal strength and noise level. The challenge of communications engineers is to find ways of packaging up signals so that they can get as close to Shannon's limit as possible.

2. In 2004, theorist Samir Mathur and colleagues at Ohio State University surprised physicists by suggesting information survives in the chaos of a black hole. The evidence comes from so-called superstring theory, according to which all particles are vibrations of multidimensional "strings". Though very tangled up by the black hole, Mathur and colleagues claim these strings retain characteristics of the original particles.

Further reading

www.lucent.com/minds/infotheo

http://researchnews.osu.edu/archive/fuzzball.htm

http://cm.bell-labs.com/cm/ms/what/shannonday/

Physics from Fisher Information by Roy Frieden (Cambridge University Press, 1999).

Three Roads to Quantum Gravity by Lee Smolin (Weidenfeld and Nicolson, 2001).

6
GM Crops

IN A NUTSHELL

The creation of genetically modified (GM) crops is one of today's most controversial scientific issues. For millennia the traits of plants and animals have been altered through cross breeding. Only since the 1940s has it become clear that this randomly changes the DNA inside cells. Genetic modification (GM) also changes the DNA – but, in principle at least, in a targeted and controlled way.

Anti-GM campaigners claim the technology could lead to a host of problems, from the emergence of "superweeds" to the development of allergies and antibiotic resistance in humans. Pro-GM scientists, in contrast, claim there is nothing new in GM, with farmers creating new varieties – such as maize from a certain type of grass – for thousands of years.

In reality, GM is a radical departure, as it focuses on just a handful of genes associated with specific traits, while conventional crop-breeding brings together huge numbers of genes with unknown consequences. Scientists claim that GM crops should be much less likely to spring nasty surprises.

Yet to date GM crops have largely benefited only agrochemical companies and farmers. The second-generation GM crops may offer direct consumer benefits such as improved nutritional value.

While getting your weekly shopping, you pick up a tin of sweetcorn – and your eye is caught by something in the list of ingredients: "Genetically modified". What do you do?

If results of a poll published recently by the Co-op supermarket group is any guide, you will put the tin straight back on the shelves: four out of five people said they would not knowingly buy GM food.

Whether that is true will soon become clear, as new EU regulations introduced in April 2004 compel food manufacturers to reveal whether their produce contains genetically modified (GM) ingredients.

The Co-op pre-empted consumer resistance by announcing that its own brand produce would be GM-free. As Britain's biggest farmer, the company also declared it would not

TIMELINE

9500 BC Central American farmers breed variety of squash which becomes the modern-day pumpkin; they later develop improved cotton, beans and maize.

1879 US botanist William James Beal at Michigan State University develops first scientific crosses of maize, and boosts yields by up to 50 per cent.

1920s Scientists find that exposing seeds to X-rays leads to creation of mutants, and opens up "mutation breeding" to improve crops such as oats and beans.

1944 Oswald Avery and colleagues at the Rockefeller Institute, New York, produce first hard evidence that DNA contains the long-sought "genes" of life.

1966 Golden Promise, a "conventionally bred" barley created using radiation from a reactor at the UK Atomic Energy Research Establishment, goes on sale.

1983 Jeff Schell and colleagues at Rijksuniversiteit in Ghent, Belgium, publish report on creation of first GM plant, an antibiotic-resistant form of tobacco.

1994 The first GM fresh food goes on sale in the US: Calgene's Flavr-Savr tomato, which its makers claimed stayed fresher longer. Overpriced, it fails commercially.

1998 Monsanto launches a major public awareness campaign in Europe, which prompts protests from anti-GM campaigners and puts the debate in the public arena.

1999 Scientists at Cornell University show that pollen from GM maize is lethal to Monarch butterflies in lab tests (field trials later refute the findings).

2003 Results of the UK field trials are published, and show that the cultivation of one GM crop – a form of maize – was more benign than the normal variety.

2004 In March the UK government gives tentative approval for planting of GM maize but its makers, Bayer Crop Science, decides not to go ahead.

use GM animal feed, or grow GM crops on its land.

The company insisted that its decision was dictated by current scientific knowledge and consumer opinion. Cynics, however, might see it as an attempt to play catch-up in a PR battle with larger rivals like Sainsbury's and Waitrose, who took similar decisions years earlier. And to many scientists, the decision flies in the face of all the evidence about the risks and benefits of GM technology.

It was, in short, yet another outbreak of contradictory arguments of the kind that have dogged the GM issue since it first made headlines worldwide in 1998. A public awareness campaign by the US agrochemicals company Monsanto prompted outrage among campaigners by claiming genetic modification of crops was nothing new.

In one sense the company was right: its awareness campaign simply brought attention to a technique for creating crops that had already been around for 15 years. But when the public learned how GM crops were being created, it did not like what it heard. It seemed that scientists were meddling with Nature, and seeking to put the results on everyone's plates.

Advocates of GM have long argued that humankind has been genetically modifying crops for millennia. Archaeological and DNA evidence suggests that farmers in Central America succeeded in creating a new form of squash akin to today's pumpkin around 9500 years ago, and later bred a mix of maize and teosinte grass. The effects of such cross-breeding was sometimes dramatic: in 1879 William James Beal at Michigan State University developed cross-breeds of maize with 50 per cent higher yield.

Yet the methods in conventional crop breeding are very different from those used in GM. Even today, conventional methods are incredibly inefficient, requiring vast numbers of plants to stand much chance of transferring traits. Worse still, a host of unknown traits can also be transferred at the same time – with sometimes serious consequences.

The potential health risks of cultivating and eating GM crops have been a central theme of the GM debate. Defenders of the technology point out, however, that because only a few genes are usually involved, the risk of unexpected side effects is much lower with GM crops than with conventionally bred varieties, where vast numbers of unknown genes are combined. Ironically, some conventionally bred crops have been at the centre of precisely the kind of scare so often associated with GM.

In 1964 a traditionally bred variety of potato called Lenape was launched, its makers claiming it made better crisps. Tests revealed, however, that the new variety also contained dangerously high levels of solanine – the toxin in Deadly Nightshade – and the crop had to be abandoned. The same problems emerged in 1995 with Magnum Bonum, a Swedish variety of conventionally bred potato.

In the late 1980s, an insect-resistant form of celery released in the US was found to contain high levels of light-activated carcinogens called psoralens which also caused severe skin reactions among workers who handled the crop once it had been harvested.

It is against this background that scientists battled to improve both the speed and quality of plant breeding. To boost the range of mutant varieties available, scientists experimented with X-rays and later multagenic chemicals

JARGON BUSTER

Marker gene: A gene used solely to reveal which GM plants have successfully acquired a new trait. The most widely used marker genes are for antibiotic resistance, which allowed plants to survive treatment with antibiotics if they had been correctly modified. Concern about antibiotic resistance being transferred to humans has led to such markers being phased out.

Terminator gene: A joint project between the US Department of Agriculture and Delta and Pine Land, the terminator gene was designed to mimic a property of many conventional hybrid plants, preventing seeds from GM crops being used more than once, thus keeping quality and yields high. Despite the advantages, the resulting controversy killed off the project in 1999.

Bt toxin: Short for Bacillus thuringiensis toxin, a protein lethal to certain insects that destroy crops such as cotton and maize. By inserting the gene responsible for creating the toxin into plants, they acquire their own insecticide, drastically reducing the need for chemical sprays.

Agrobacterium tumefaciens: A bacterium found in soil which infects plants with its genetic material, triggering disease. By removing the genes responsible for disease, scientists used A. tumefaciens to carry genes for desired traits, producing GM varieties of crops such as potatoes, tomatoes and soya beans.

Organic farming: Begun in 1940s Britain, organic farming focuses on the health of the soil in which crops grow. By avoiding intensive field use and artificial fertilisers, organic farmers claim to produce better, tastier crops – albeit at a higher cost. Many experts believe the method is too expensive for mass agriculture, however, and that GM crops offer a better bet.

The strange story of atomic barley

To speed the development of new crops, scientists in the 1950s began using a method of genetic modification that would cause banner headlines today: nuclear radiation. By exposing seeds to gamma rays from reactors, scientists created hundreds of mutant varieties of standard crops.

The most famous is Golden Promise, a mutant form of Maythorpe barley created at the Atomic Energy Research Establishment at Harwell, Oxfordshire. First sold in 1966, it is shorter than standard barley but with good malting qualities and a higher yield. This made it very attractive for brewing, and it is still a key ingredient of some of Scotland's finest single malt whisky.

and even nuclear radiation to produce new "natural" crops still in use today. Even so, the success rate remained pitifully low.

The first glimmerings of a new approach emerged in the 1940s, as scientists unravelled the true nature of genes, the long-sought carriers of biological information that allowed traits to be passed down the generations. DNA was identified as the key molecule, with genes turning out to be sequences of chemicals called "bases" strung along the helix-shaped molecule like pearls on a necklace.

These discoveries opened up a radical new possibility for crop breeding. Instead of blindly combining all the genes from two plants in the hope that, say, the genes controlling yield would be transferred, scientists pondered the possibility of identifying the handful of genes involved in the trait, and transferring just those to another plant.

As well as being far less haphazard and inefficient, this tightly focused genetic modification seemed less likely to produce unwanted side effects, such as toxicity or allergic responses. The challenge was to find a way of getting the required genes into the genome of the target plant.

By the early 1980s, attention focused on a bacterium called *Agrobacterium tumefaciens*, which has evolved a way of smuggling some of its own genes into plants in a package known as a plasmid.[1] In 1983, scientists from the Rijksuniversiteit in Ghent, Belgium, announced they had succeeded in using *A. tumefaciens* to insert a gene for antibiotic resistance into tobacco plants.

In creating the world's first GM plant, the Belgian team had also developed a key technique needed if GM was ever to rival conventional crop breeding. By adding an antibiotic resistance "marker gene" to genes for traits like improved yield, scientists could discover which plants had been successfully modified simply by treating them with antibiotics that affect growth – and seeing which ones continued to thrive.

The creation of GM tobacco sparked an explosion of research which initially focused on simple, single-gene properties such as herbicide resistance – which allowed crops to thrive when dowsed with chemicals lethal to surrounding weeds – and pest resistance, with crops being given a gene making them toxic to insects that try to eat them, thus eliminating the need to spray the whole field with chemicals.

But at the same time, eco-activists became concerned by what they saw as tinkering with Nature simply to boost profits for the farming and agrochemical industries.[2] Much of their concern focused on attempts by scientists to give crops radically new traits using genes taken from utterly unrelated species such as fish. Dubbed "Frankenfood" by campaigners, such produce seemed to push crop breeding beyond the bounds of knowledge. Others raised the spectre of "superweeds" created by genes spreading from herbicide-resistant GM crops to other plants, and of a wave of allergies and antibiotic resistance among humans triggered by eating GM crops.

At first public concern was muted. In 1994, US-based Calgene launched the first GM fresh food, the Flavr-Savr tomato, said to remain fresher longer. Reports circulated that the tomato had been created using a gene taken from a fish; in fact, its improved shelf-life was due to a modified version of a standard tomato gene. This first attempt to use GM technology to benefit consumers rather than farmers soon soured, however, as the tomato offered little real improvement, but cost twice as much.

Ironically, it was an attempt by a leading agrochemicals company to reassure consumers that turned the GM debate into a firestorm. In June 1998, Monsanto launched a £1 million advertising campaign in the UK to "encourage a positive understanding of food biotechnology". The adverts implied that GM tomatoes and potatoes had been approved, and that Monsanto had been carrying out safety tests on GM plants for 20 years. A flood of complaints followed, and the company was censured by the Advertising Standards Authority for making misleading claims. Monsanto conceded it had made a PR blunder, and set about trying to change its image. By then, however, the GM issue had become one of the most contentious scientific debates of recent times – fuelled by a spate of incidents apparently confirming the fears of environmentalists.

In 1999, scientists from Cornell University, New York, reported that in

Superweeds

Many GM crops have been created using genes that make them resistant to herbicides. This has raised the spectre of these genes finding their way into other plants, creating species of unstoppable "superweeds" against which weedkillers are powerless to act. Advocates of GM technology argue that the genetic differences between crops and weeds reduce the risk of gene transfer, adding that herbicide resistance has been created in conventionally bred crops for many years, without causing major problems. Even so, scientists concede the risk is not zero – though they insist some herbicide will always be available to zap the weedy consequences.

laboratory tests pollen from Bt maize – genetically modified to produce the *Bacillus thuringiensis* toxin, to kill corn borers – was lethal to larvae of the famous Monarch butterfly. The following year, a farmer in Alberta, Canada, reported that ordinary canola plants had acquired resistance to three weedkillers from nearby herbicide resistant GM canola – heightening fears about "superweeds". Then in September 2000, a form of Bt maize never intended for human consumption was found in taco shells made by a major US fast-food chain. Known as StarLink, the maize had been approved only for animal feed following concern about possible allergic reactions in humans.[3]

Reports of allergic reactions among dozens of people began to emerge, and Aventis CropScience, which marketed StarLink, immediately ceased production and began a £60 million recall programme. Farmers sued the company for allegedly failing to warn them about the restrictions.

Yet tests by US government scientists later found no evidence that StarLink had been responsible for the allergic responses. It was a similar story with the scare over Monarch butterflies: subsequent real-life field studies of the effect of Bt maize showed no adverse effects. And the "superweed" incident in Alberta turned out to be the result of growing conventional canola far too close to the GM variety – and only other canola plants, rather than real weeds, had taken up the genes in any case.

All this made little difference, however: the damage to the image of GM crops had been done. When the UK Government set up its GM Nation debate in 2003, its poll of 40,000 people showed that most never wanted to see GM crops grown in Britain.

They may yet get their wish. In October 2003, scientists in Britain published the results of the world's largest ever field trials of several GM crops, and concluded that a herbicide-resistant GM maize was better for the environment than its conventional equivalent. As a result, in March 2004 the Government approved the crop for planting. Days later, the makers of the crop, Bayer CropScience, pulled out, saying that the tight regulations imposed had made it uneconomic.

It now seems that no GM crops will be planted in the UK for at least the next few years. Despite this, GM business is thriving worldwide – as evidence emerges that GM crops do bring the benefits claimed by its makers, such as drastically reduced insecticide use and higher yields. Around twenty countries now grow GM crops, which accounts for around 13 per cent of the global commercial seed market.

What is still lacking, however, is a single GM product with direct benefits for consumers. In principle, genetic modification can produce food that tastes better, has better nutritional content, or even combats health problems like diabetes. Yet in many cases scientists have yet to find the plant genes responsible, let alone ways of modifying them. Golden Rice, genetically modified to boost its vitamin A content, is undergoing tests in the Far East, but is still years from reaching the market.[4]

Many scientists remain convinced that GM methods are the future. They argue that conventional breeding techniques are not just less efficient, but also riskier – in that genes with unknown effects can be transferred along with those for the desired trait. Such crops are also planted and marketed with no real attempt to study their impact on the environment – or us. In short, they say, GM is the way we should always have bred new crops.

Sceptics, however, insist we still don't know enough about the long-term effects of planting or consuming GM crops.

The big question now is whether the public can ever learn to trust the science behind GM – or the companies who have so far promised consumers so much, yet delivered so little.

Notes

1. In most organisms, the complete DNA molecule is divided up into "packages" called chromosomes. Bacteria have other stashes of DNA that lie outside these packages in the form of so-called plasmids. These are used to transfer bacterial genes to other bacteria and organisms – in other words, to infect them. As such, plasmids are natural DNA "vectors" which can be used to carry specific genes into a target organism. The first GM crops of tobacco and cotton were created using plasmids from the soil-dwelling bacterium *Agrobacterium tumefaciens*, which were modified to switch off the genes responsible for plant disease, and instead carry genes for the desired traits into the plants. While plasmids work with some plants, they fail with many others, including cereals.

2. During the 1960s, environmentalists led by the US biologist Rachel Carson, author of *Silent Spring*, raised concern about insecticides then in use, such as DDT. Arguing that these industrial compounds were damaging the ecosystem, Carson called for more natural alternatives – including the Bt toxin, which is now incorporated into GM varieties of crop. Advocates of GM insist that their approach reduces the need to spray vast areas with insecticide, minimising the impact on the environment – and those applying them.

3. In the mid-1990s, Plant Genetic Systems of Belgium created a new form of maize using a special variety of the so-called Bt gene used in other GM crops to make them resistant to insects. Known as StarLink, the crop included a gene for a toxin lethal to pests that tried to burrow into or eat the crop. In applying to the US Environmental Protection Agency for a planting licence, its makers warned that the toxin had similarities to proteins responsible for allergies in humans. The EPA responded by approving the crop – but only for use in animal feed. It was a restriction that proved impossible to maintain: StarLink became mixed up with non-GM varieties in farm elevators and vehicles – and ended up in human food, triggering a major scandal.

4. The result of a *pro bono* collaboration between scientists and the GM industry, Golden Rice is a genetically modified version of this staple Asian food that has a higher content of vitamin A – a lack of which threatens the health of 400 million people worldwide. It is now undergoing tests at the International Rice Research Institute (IRRI) in the Philippines, but is still several years from approval and general release.

Further reading

Eat Your Genes by Stephen Nottingham (Zed Books, 1998)

A Consumer's Guide to Genetically Modified Food by Alan McHughen (OUP, 2000)

Safe Food by Marion Nestle (University of California Press, 2003)

www.bbc.co.uk/science/genes/gm_genie/

www.colostate.edu/programs/lifesciences/TransgenicCrops

LIFE

7
Out of Africa

IN A NUTSHELL

In 1871, Charles Darwin argued that the similarities between apes and humans pointed to Africa as the cradle of modern humans, *Homo sapiens*. For many years, this seemed implausible, with fossil remains of human-like creatures turning up only in Europe and Asia. Early studies of blood groups also revealed similarities between humans and African apes, yet not until the 1920s and the discovery of *Australopithecus*, a kind of an ape-like prehistoric human, did attention focus on Africa. By the mid-1980s, many fossils of human-like creatures had been found, but argument raged over their role in the evolution of modern humans. Some claimed that *H. sapiens* evolved in many different sites across the world from more primitive *H. erectus*, which had left Africa long ago. Others insisted that *H. sapiens* had first emerged in Africa, and had walked out of the continent relatively recently.

This "Out of Africa" view is now supported by comparing DNA in living humans, which suggests *H. sapiens* existed around 120–220,000 years ago in Africa. A group of perhaps just 10,000 of them then left the continent around 50–100,000 years ago, perhaps because of climate change. There they encountered the descendants of *H. erectus* in Asia and Europe, including Neanderthals. After a brief period of co-existence, *H. sapiens* went on to dominate the world.

With his piercing eyes and powerful physique, the Ethiopian man who appeared before the world's media in June 2004 was clearly not someone to be trifled with. His celebrity status had even put him on the cover of *Nature*, the world's leading science journal. That is an accolade denied even to

most Nobel Prize winners. It is even more unusual for someone who has been dead for over 150,000 years. He is Herto Man, whose remains were discovered in Ethiopa's Afar valley, 230 km north-east of Addis Ababa. The first part of his name comes from the nearby village of Herto; it is the second

TIMELINE

1856   Quarrymen working near Dusseldorf, Germany, uncover bones of what is now called Neanderthal Man.

1871   Charles Darwin predicts in *The Descent of Man* that the origins of modern humans will be found in Africa.

1891   Dutch physician Eugene Dubois finds remnants of an early human in Java, now known to be *Homo erectus*.

1904   Professor George Nuttall of Cambridge University uses blood groups to uncover first hints of link between humans and African apes.

1924   Raymond Dart identifies ape-like human fossil *Australopithecus africanus*, resurrecting Darwin's Out of Africa theory.

1987   Allan Wilson of University of California and colleagues use "Mitochondrial Eve" idea to argue that humans came from Africa less than 290,000 years ago.

1992   University of Utah researchers claim DNA suggests modern humans are descended from just 10,000 who left Africa 50–100,000 years ago.

1994   Using new dating methods, scientists at Berkeley show *H. erectus* reached Java around 1.8 million years ago.

1997   Mark Stoneking of Penn State University and colleagues analyse DNA from Neanderthals and find no evidence of genetic link to modern humans.

1998   Professor Stanley Ambrose of University of Illinois suggests humans left Africa to escape climatic change triggered by volcanic eruption 71,000 years ago.

2000   Ulf Gyllensten and colleagues at the University of Uppsala, Sweden, reveal DNA evidence points to modern humans originating in Africa 170,000 years ago.

part of his name that explains his new found celebrity. For scientists believe Herto Man is the oldest known member of the species which now dominates the planet: modern humans, or *Homo sapiens* – "Knowing Man".

The great antiquity of Herto Man, combined with the location of his last resting place, are being hailed as key pieces of a jigsaw puzzle that began almost 150 years ago in a German quarry. Yet only in the last few years has the full picture begun to take shape. What it shows is that the apparently huge diversity of today's humans, from Scandinavians to Sumatrans, Inuits to Indians, is just an illusion. Ultimately, we are all descended from a single group of people from one part of the world: Africa.

This would come as no surprise to Charles Darwin, the father of evolution. As long ago as 1871, he argued in his book *The Descent of Man* that all the races of humans were just varieties of a single species, with Africa – the home of the world's great apes – as the most plausible birthplace of modern humans.

Today, with spectacular human fossils like Herto Man routinely emerging from African sites, Darwin's claim may seem unremarkable. Yet at the time he was writing, all the evidence pointed to a birthplace much closer to home: Europe. Part of a fossilised human-like skull had been found in southern Germany around 1700, though its significance was not recognised for over a century. In 1856 quarrymen working in a cave in the Neander Valley near Dusseldorf found bones belonging to some kind of early human, but dumped them as worthless. Retrieved by a local teacher some weeks later, the bones were hailed by some academics as the remains of an extinct species of human: *Homo neanderthalensis* – "Man from the Neander".

In what was to become almost a ritual following the discovery of any supposedly human fossil, the bones of *H. neanderthalensis* prompted enormous controversy. Some anatomists agreed that they had come from members of a previously unknown species of human, while others pointed to the similarity of skull size with modern humans, insisting the bones merely came from an ancient example of *H. sapiens*. As more bones emerged at other sites, some anatomists tried to settle the argument by highlighting supposed differences between Neanderthals and modern humans, such as their "knuckle-grazing" walk (which in the 1950s was proved to be a myth, based on studies of a Neanderthal with severe arthritis).[1]

When the next major piece of the jigsaw of human origins emerged, it too failed to fit Darwin's "Out of Africa" theory. In 1891 the Dutch physician and fossil-hunter Eugene Dubois found fossilised fragments of a human-like creature in Java. With its much smaller skull, "Java Man" was unlike either modern humans or Neanderthals. Subsequently renamed *Homo erectus* – "Upright Man" – Dubois's find was regarded as another genuinely new species of human. Ironically, its resting place, thousands of miles from Darwin's putative cradle of humanity, would turn out to be a key part of the African story of *H. sapiens*.

Not until 1904, over 20 years after Darwin's death, did fresh scientific evidence emerge to back his view of human origins. George Nuttall, a bacteriologist at Cambridge University, had become intrigued by the newly identified blood groups of humans, and carried out tests comparing human blood to that of other creatures. To his astonishment, Nuttall found striking similarities between the blood of humans and that of apes from Africa. The implications were no less astonishing. Despite appearances – and

JARGON BUSTER

Multiregional Hypothesis: The principal rival to the Out of Africa theory as an explanation to the origins of modern humans. It claims that members of *H. erectus* evolved into *H. sapiens* wherever they settled after leaving Africa around 1.8 million years ago, in some cases after passing through a Neanderthal phase.

Mitochondrial DNA (mtDNA): DNA extracted from the energy generating parts of human cells known as mitochondria. As large numbers of mitochondria exist in each cell, their DNA is easier to extract from ancient fragments. The DNA is also more varied between individuals, and is inherited only via females, making any changes easier to interpret.

Missing link: Originally held to be the creature that bridged the evolutionary gap between chimps and humans. It is now known that humans are not descended from chimps, but that both are descended from an unknown common ancestor at least 5 million years ago – which has become the real "missing link".

Homo erectus: The species of human that left Africa and spread into Asia around 1.8 million years ago. According to the Out of Africa hypothesis, those that remained in Africa eventually became *Homo sapiens*, and replaced *H. erectus* and its descendants in Europe and Asia. According to the Multiregional Hypothesis, *H. erectus* evolved into *H. sapiens* at various sites around the world.

Are there really so many human species?

Every few years, another team of fossil hunters hits the headlines with the discovery of another new species of human. Since the identification of Neanderthals around 150 years ago, evidence has been found for around a dozen different human species. Or has it? Most of the claims are based on examination of a handful of bones from just a few individuals. This should immediately raise concern about inadequate sampling – after all, modern humans show startling diversity in size, despite belonging to the same species. These suspicions have now been made quantitative by the work of Professor Maciej Henneberg of the University of Adelaide. In 2004 he published a detailed statistical analysis of over two hundred specimens of human-like fossils from four million year-old Australopithecines to Neolithic modern humans. His results confirmed that there has been substantial evolution in both skull sizes and body weight over that time. But crucially, he also found there was no evidence for the fossils being anything other than a *single* species which had grown bigger and smarter over time.

the fossil evidence up to that time – modern humans shared common ancestry with creatures from Africa.

Unknown to Nuttall, his discovery was the first glimpse of the power of DNA to probe questions of human origins. Years later, scientists would refine his methods to give impressively detailed insights into the events that led up to the origins of *H. sapiens*. For several decades after his pioneering work, however, the focus of attention would switch back to fossil evidence.

The prospect of finding human remains in Africa had led fossil hunters there as early as 1912, but with few clues about where or what to look for, they came away empty-handed. Bizarrely, the breakthrough came in the form of a fossilised skull sitting on the mantlepiece of a quarry manager in Taung, South Africa. It was spotted one day in 1924 by a student of Raymond Dart, an Australian professor of anatomy at the University of Witwatersrand in Johannesburg. He quickly identified the skull as that of an extinct baboon – and began wondering what else had been found at the quarry.

Shortly afterwards, Dart received two boxes of material dug up by the quarry workers. The first contained nothing of value, but in the second he found fragments of a skull-shaped fossil. Their small size suggested an ape-like creature, but the jaws, teeth and other anatomical details were unlike those possessed even by chimpanzees, our closest living relatives. Dart recalled Darwin's old Out of Africa theory, and wondered if he was looking at the remains of the creature from which both humans and chimpanzees had evolved – the fabled "Missing Link".

Within a few months, Dart's discovery was making worldwide headlines. He named the creature *Australopithecus africanus* – "Southern African ape" – and immediately ran into a barrage of criticism. Leading scientists insisted he was pushing the evidence much too far, and that the skull was simply a fossilised ape's head. Others rejected the implication of an African origin for humans on essentially racist grounds: to them, it was inconceivable that modern humans could have emerged from the homeland of slaves.

The significance of Dart's discovery was rejected until the 1940s, by which time many more specimens of *Australopithecus* had emerged to support his view that they were a distant ancestor of humans. Africa now became the focus of the quest for human origins. Ever more specimens of human-like creatures began to emerge, with names like *Australopithecus afarensis*, *A. boisei* and *Homo habilis*. Yet the emerging picture of human origins was, if anything, becoming less clear.

Arguments broke out between rival teams of scientists over the relationship between all these different creatures. Which ones were direct descendants of modern humans, and which just evolutionary dead ends? And how could the African origins of modern humans ever be proved?

A key part of the puzzle fell into place in 1960. While working in the Olduvai Gorge of Tanzania, the Kenyan fossil hunters Louis and Mary Leakey found part of a skull belonging to *H. erectus* – the same human species Dubois had found thousands of miles away in Java. The discovery formed a crucial link between prehistoric humans in Africa and the rest of the world. By the mid-1980s, fossil hunters had found many more examples of *H. erectus* in Africa, including the astonishingly complete skeleton of a child, "Turkana Boy" found by Richard Leakey (Louis's son), in 1984.

Most importantly, a pattern began to emerge. Radioactive dating showed that the African *H. erectus* fossils were up to 1.8 million years old – far older than most of the specimens found elsewhere. This pointed to *H. erectus* having left Africa well over a million years ago, walking out into Asia, the Middle East and Europe – and then presumably evolving into Neanderthals and even *H. sapiens* in many different locations.

While appealingly simple, this so-called "multi-regional" view of human evolution was hardly compelling. Modern humans may actually have evolved from the *H. erectus* which remained in Africa, leaving the continent in a second exodus long after the first. Over a century after Darwin's original proposal, this became the central claim of the Out of Africa theory: that *modern* humans had first

Humans may have taken the scenic route out of Africa

While most scientists now accept the Out of Africa Theory, the route taken by the first humans to leave the continent remains controversial. Until recently, it was generally assumed that they took a northerly route, walking into the Middle East and then spreading out from there. However, mtDNA analysis now suggests the exodus may have proceeded via a more southerly route, along the African coastline. In 2005, an international team of researchers announced that an isolated group living in Malaysia appeared to be the descendants of humans who left Africa around 65,000 years ago. According to the researchers, climatic change underway at the time would have made a southerly route easier – not least because Europe was then in the grip of an Ice Age. The genetic evidence suggests perhaps as few as several hundred individuals went first to India, then Southeast Asia and Australasia. If correct, this would explain why humans appear to have reached Australia around 50,000 years ago, while the oldest human remains in Europe – a jawbone found in Romania – are only around 35,000 years old.

emerged in Africa, and left the continent long after *H. erectus*, ultimately replacing them across the globe.

Proving it was a different matter, however; there was always the possibility that fossil hunters had simply missed some vital clue. Some other source of evidence was needed, and during the 1960s it began to take shape in the laboratory of biochemist Allan Wilson and colleagues at the University of California.

They resurrected the basic idea behind George Nuttall's blood tests: that today's humans carry clues to their origins in their cells. As ever, their early findings provoked enormous controversy. By studying the blood proteins of humans and apes, they claimed the two had parted company around 5 million years ago – far more recently than the fossil evidence suggested. Despite a storm of protest, later fossil evidence backed Wilson's chronology.

Encouraged by this success, the team went on to devise another technique, based on so-called mitochondrial DNA (mtDNA) found in living human cells. By comparing the mtDNA of modern humans from different parts of the world, Wilson hoped to gauge just how recently we all shared a common ancestor. The team found relatively little difference between the mtDNA of modern humans, consistent with *H. sapiens* having emerged less than 290,000 years ago. They also found that the biggest differences were among Africans – just as expected if they are the oldest members of *H. sapiens*.

Wilson's findings made headlines around the world, and spurred DNA-based research by many others. The results provide a stunningly detailed picture of the origin of modern humans. First, they point to Africa as the birthplace of *H. sapiens* around 120–220,000 years ago. Second, they suggest that around 50,000 to 100,000 years ago, perhaps as few as 500 members of the species walked out of Africa to become the ancestors of all non-African races. Arriving first in Asia and later Europe, they encountered the descendants of *H. erectus* who had left Africa around 1.8 million years earlier. After a brief period of co-existence with the Neanderthals, by around 35,000 years ago *H. sapiens* had become the dominant species it is today.

Many puzzles still remain. For example, what drove *H. sapiens* to leave Africa? One possibility is climate change: a volcanic eruption took place on the island of Sumatra around 71,000 years ago – just when DNA evidence suggests the exodus took place.[2] Then there is the mystery of the Neanderthals, which DNA evidence suggests were unrelated to modern humans. Why did they die out?

The fossil hunters are continuing their search for answers. For now, it seems all but certain that the most successful creatures the world has ever seen really did come, as Darwin claimed over a century ago, Out of Africa.

Notes

1. Palaeoanthropology has had more than its fair share of embarrassing revelations over the years – symptomatic of the tendency of some fossil hunters to read far too much into far too little evidence. The most notorious example is the Piltdown Man hoax, which centred on the skull and jawbone unearthed by amateur geologist Charles Dawson near Piltdown Common, Sussex, UK, around 1912. The find was quickly accepted as genuine because it seemed to fit so well with prevailing ideas about the evolutionary links between humans and apes. Not until the 1950s did it emerge that it had been constructed out of the skull of a modern human plus the jawbone of a newly dead orang-utan. Perhaps the most egregious example of evidence being pushed too far dates back to 1922, when the American fossil expert Henry Fairfield Osborn claimed to have found "Hesperopithecus", the first anthropoid ape in America. Osborn based his claim on his discovery of a single fossilised tooth – later shown to have come from an extinct pig.

2. Climate change is regarded as among the greatest threats now facing modern humans. Yet very recently evidence has emerged to suggest that it may actually have played a key role in giving us our most impressive attribute: our brains.

Oddly enough, the sheer size of our brains is something of an evolutionary puzzle. While it clearly pays to be smart, our brains come with a hefty price tag in terms of energy demand. Despite weighing only around 1.4 kg – just a few per cent of our total body weight – our brains consume around 20 per cent of our total energy intake.

Many other organisms have evolved similarly profligate thinking machines, so clearly there is some major evolutionary benefit to be gained from having a decent-sized brain. But what is it? In 2005, a team led by Dr Daniel Sol of the Autonomic University of Barcelona showed that the answer may lie in climate change. In a paper in the *Proceedings of the National Academy of Sciences*, they showed that data on the success of bird species introduced into new environments supported a link between brain sizes and the ability to develop new skills and survive. The birds with bigger brains were better able to spot new sources of food and exploit them. According to the team, the findings provide strong evidence to support the view that large brains evolved to deal with environmental change.

Does this explain why humans have such huge brains? We are certainly the smartest creatures this planet has ever seen, and we have proved very capable at coping with the climatic upheaval of the last few million years. Some may worry about whether we shall cope with the effects of global warming, but history suggests that the 1.4 kg mass of squidgy stuff on top of our necks will see us through.

Further reading

The Origin of Humankind by Richard Leakey (Weidenfeld & Nicolson, 1994)

African Exodus by Chris Stringer and Robin McKie (Owl Books, 1998)

Missing Links: The Hunt for Earliest Man by John Reader (Penguin, 1988)

8
Nature versus Nurture

IN A NUTSHELL

Are we primarily shaped by our experiences and upbringing, or do our genes dictate what we can become? For centuries philosophers and social reformers tried to solve the riddle of "nature versus nurture", but only relatively recently has scientific evidence been brought into the debate. All too often, however, the evidence has cast more light on the beliefs of researchers than the roles of genes and environment. A century ago, the evidence was said to prove the dominance of genetic factors – and was used by eugenicists to justify mass sterilisation campaigns among the genetically "unfit". By the 1930s, the pendulum had swung back towards the nurture camp, with behaviourists insisting humans have no innate traits.

Only in the last decade or so has a more balanced view emerged, according to which humans are shaped partly by their genes, partly by their environment – and partly by sheer happenstance.

For the tabloids, it was an irresistible story: a scientific explanation of the bed-hopping antics that fill their pages day after day. "Were you born to stray?" asked the headline in a July 2004 *Daily Mail*, above a report of research by scientists at St Thomas's Hospital, London, suggesting that around a quarter of women carry an "infidelity gene" making them more likely to have affairs.

Professor Tim Spector and his colleagues at the Twin Research Unit had compared the personal records of thousands of female twins on their database, and found that identical twin sisters – who share the same genes –

showed a striking similarity in their rate of infidelity.

The implication was obvious: a woman who keeps having flings just can't help it – it's in her genes. That at least was how it was seen by the tabloids, who spiced up their reports with examples of celebrities notorious for their philandering. Only towards the end of the stories did the researchers' caveats finally appear – that genes do not compel behaviour, and that infidelity is likely to depend on a host of other factors in any case.

What the research did prove beyond all doubt was the abiding interest in the

TIMELINE

1690   English philosopher John Locke puts forward the claim that every human is born a "blank slate" and acquires characteristics through experience of life.

1864   Herbert Spencer coins the phrase "survival of the fittest", and lays foundation for application of Darwinism to society, later called Social Darwinism.

1875   English polymath Francis Galton publishes first study of twins, designed to assess relative balance of nature and nurture. He later coins the term "eugenics".

1924   American psychologist John B. Watson puts forward "behaviourism", which sees all human abilities as the result of environmental interaction rather than traits.

1931   Psychologist Winthrop Kellogg of Indiana University and his wife raise a chimp alongside their son to study effects of genes and environment on behaviour.

1935   Eugenics-led policy leads all US states to segregate people with mental disabilities, and thirty-five states introduce compulsory sterilisation; 20,000 "feeble-minded" are sterilised in California alone.

1943   British educational psychologist Cyril Burt publishes evidence from twin study that intelligence is largely determined by genes, prompting changes in UK education system.

1958   American psychologist Harry Harlow unveils results of his experiments with monkeys, showing how gene-based behaviour can override learned behaviour.

1975   Publication of *Sociobiology* by Harvard entomologist Edward O. Wilson rekindles nature versus nurture debate, with Wilson accused of genetic determinism.

1998   Judith Rich Harris publishes *The Nurture Assumption*, which debunks years of scientific claims about the roles of nature and nurture in childhood development.

2001   Study of the human genome reveals only around 30,000 genes, suggesting that outside influences play key role in activation of genes.

nature versus nurture debate. Hardly a month goes by without some research team announcing a link between genes and some or other personality trait, from risk-taking to sexual orientation, bullying to insanity. And behind the reports of the research lurks the unnerving implication that we are all in thrall to our genes.

It is a belief that has underpinned such outrages as the forcible sterilisation of "feeble-minded" people in 1930s America to the ethnic cleansing in the Balkans of the 1990s. Yet the reaction against genetic determinism has produced its own excesses. Belief that humans are "blank slates" whose future is determined entirely by their environment has led to bizarre theories of childrearing – and untold guilt among parents who blame themselves for the failings of their offspring.[1]

Now a new and more sophisticated picture is beginning to emerge: a relationship between our genes and upbringing and environment that is far more complex than the demagogues of the debate would have us believe.

Not that the pioneers of the debate saw themselves as having anything other than the best intentions for society.

When the seventeenth-century English philosopher John Locke stated the case for the "blank slate" view of human behaviour, he believed he was striking a blow against such repressive concepts as original sin and the divine right of kings. If humans were all born equal, then everyone could, and should, have as much right to life, liberty and the pursuit of happiness as anyone else – a view which impressed Thomas Jefferson, principal architect of the US Declaration of Independence. In the same way, when the Victorian intellectuals Herbert

Spencer and Francis Galton tied Darwin's theory of evolution to the study of human society, they believed they were acting in the public good.[2]

Some of their contemporaries could see the dangers lurking in Spencer's famously pithy summation of evolution as "the survival of the fittest" and Galton's concept of eugenics – the systematic "improvement" of the human race through selective breeding. Even so, many intellectuals simply brushed aside such qualms, believing that the facts spoke for themselves. As early as 1865, Galton had published a study of the eminence of children from notable families, and concluded their success rate was 240 times that of the offspring of the general public. A decade later, Galton followed this up with the first of what became a staple of the nature versus nurture debate: a comparison of identical twins. Finding so many similarities between such twins throughout their lives, Galton believed there was no debate to be had: nature clearly prevails over nurture – and selective breeding was the way forward for society.

It was a view shared by many of those who attended the First International Congress of Eugenics in London in 1912 – including Leonard Darwin, son of Charles. As president of the Eugenics Society, he warned darkly of the threat to future generations of allowing the "unfit amongst men" to breed. Yet even at the conference, dissenting voices could be heard. In his address to the Congress, Arthur Balfour, the former British prime minister, expressed his concern that the whole question of heredity was far more complex than scientists believed – and warned of the hijacking of eugenics by zealots keen to foist their own views on society.

Balfour's views proved all too prescient in both respects. In the United States, the concept of eugenics spread like wildfire: a year after Darwin's speech, sixteen states had laws permitting compulsory sterilisation of the "feeble-minded". The campaign was watched by eugenicists

JARGON BUSTER

Behaviourism: The belief that behaviour is almost entirely due to learning from experiences gained in particular environments. First proposed by the American psychologist John Watson in 1913, behaviourism led to the notion that all behaviour can be altered using rewards and punishment.

Heritability: A measure of the degree to which differences in a particular trait are due to genetic factors. For example, the heritability of IQ is around 0.5, implying that genetic effects account for around 50 per cent of the variability in IQ between people. Similar figures apply to personality traits and success in life.

Eugenics: Term first coined by Sir Francis Galton from the Greek "eugenes" meaning "good in birth", eugenics involves the deliberate use of selective breeding among humans to improve the genetic composition of the population, with high achievers encouraged to have more children.

Twins studies: The use of twins – usually identical – to show the relative importance of genetic and environmental influences on a particular trait, such as height, IQ and personality type.

The blank slate: The view that humans are born with no innate traits, so that their behaviour, abilities and personalities are entirely the result of interaction with their environment.

John Locke – the man

One of the most influential of all British philosophers, Locke was born in 1632 and was educated at Christ Church, Oxford. While there he developed a fascination for medicine and science, encouraged by fellow Oxonian Robert Boyle. His studies also encompassed philosophy, and questions about the nature of the human mind and perceptions of reality. In his most famous book, *Essay Concerning Human Understanding*, Locke argued that the human mind starts out entirely blank, and only acquires knowledge about the world via experience. This "Empiricist" view led him to conclude that there are limits to both the extent and certainty of knowledge – a strikingly modern argument. His ideas about the blank human mind, in contrast, are now untenable.

in Germany, whose policies were introduced within months of the Nazis coming to power in 1933. They began with sterilisation of thousands with traits like schizophrenia and ended with the slaughter of millions in death-camps like Auschwitz.

The defeat of the Nazis, and revulsion at their policies, led the nature versus nurture debate to lurch back to the "blank slate" view of human behaviour. Once again, proponents were able to call on apparently solid scientific research to back their views. And once again, their conclusions rode roughshod over the true complexity of the situation.

During the 1920s, the American psychologist John B. Watson had proclaimed all talk of traits and instincts as beyond quantification and thus meaningless. Instead, he called for a focus on how humans behave in response to the world around them. This, he argued, would prove that all humans have the capacity to achieve anything.

Watson and his acolytes gathered a wealth of evidence to back their claims – some of it decidedly eccentric. Winthrop Kellogg studied the role of environment in behaviour by raising an ape with his child, while the Harvard behaviourist B.F. Skinner put his baby daughter in a specially constructed box for several hours a day, arguing that it created an optimal environment for her development. Watson himself even "conditioned" his own child to have an irrational fear of rabbits.[3]

More bizarre still was the abject failure of behaviourists to accept that while their research was certainly consistent with the importance of nurture, it did nothing to rule out a role for other factors – including genetics. Just how important these other factors could be was graphically demonstrated in controversial experiments performed in the late 1950s at the University of Wisconsin-Madison.

Psychologist Harry Harlow separated baby monkeys from their mothers and put them in cages containing two artificial mothers. One was just a wire-frame model fitted with a feeding bottle and milk, while the other was more life-like and cuddly – but unable to supply milk. According to the behaviourists, the monkeys should have quickly learned to ignore the coldness of the wire "mother", and got on with the vital business of taking her milk. Yet Harlow found the monkeys spent most time with the cuddly but milkless mother, making only brief dashes to the wire version when hungry.

Harlow had demonstrated what most people – except behaviourists – would regard as obvious: behaviour is shaped by more than just the environment. The monkeys had an in-built instinct for what to expect from a nurturing parent, and sought it out.[4]

Harlow's evidence for innate behavioural traits came at a time when both sides of the nature versus nurture debate were having profound effects on parents and children. Behaviourists wrote childcare manuals that insisted children would become "soft" if kissed goodnight or cuddled excessively. At the same time, evidence for the influence of genetics was influencing educational policy. Studies of identical twins by the British educational psychologist Cyril Burt were said to show that intelligence is largely inherited, prompting claims that educational resources should be focused on children who showed early promise. In the UK, this led to the 1944 Education Act and the 11-plus examination, which selected 11-year-olds for grammar school education.

Years later the twins studies behind this policy came under suspicion, but by then the claim that genes are destiny had lost much of its power – at least outside academia. Parents had long recognised that despite their best efforts, their offspring often ended up with wildly different personalities that showed little sign of being "determined" either by nature or nurture.

Among academics, however, the debate rumbled on. In 1975 the Harvard ant expert Edward O. Wilson published *Sociobiology*, in which he showed that genes alone can produce spectacularly complex behaviour. But he also attempted to extend his arguments to human behaviour, and provoked a storm of controversy – not least because it seemed to resurrect genetic determinism, with its eugenicist overtones.[5] Meanwhile, scientists continued to claim evidence for genetic

Galton's experiments on eminence

Prompted partly by his own illustrious family background, Galton set about collecting hard evidence for the heritability of talent by scouring the biographies of famous people. After analysing the family histories of hundreds of such people born during the four centuries following 1453, Galton found around 8 per cent of sons of distinguished people also became eminent. For comparison, he estimated the proportion of the general population who at least succeeded in attending university, and found a figure of just 1 in 3000. Galton believed this 240-fold difference in success rate was due primarily to inherited factors – though he conceded that the offspring of eminent people may well enjoy advantages denied to the rest of the population.

influence on everything from sexual orientation to career choice.

By the mid-1990s, the academic world finally appeared to be coming round to the same conclusion as the public: that human behaviour is a mix of nature, nurture and simple happenstance. In 1998, the American psychologist Judith Rich Harris published *The Nurture Assumption*, a best-selling book which gave scientific backing for what parents had always suspected: that their parenting skills have relatively little effect on how children turn out. At the same time, studies of the human genome uncovered genes whose actions affect the ability to interact with others – making a mockery of the standard dichotomy between genetic and environmental factors.

After a century and more of claim and counterclaim, the nature versus nurture debate is finally being seen for what it is: an object-lesson in the dangers of thinking one side of any debate has a monopoly on the truth.

Notes

1. A literal translation of the Latin "tabula rasa", the blank slate captures the idea that humans are born without innate traits, their behaviour and personalities being entirely the result of interaction with their environment. In his original statement of the concept, John Locke used the metaphor of the mind as "white paper devoid of all characters, without any ideas".

2. One of the most original Victorian thinkers, Francis Galton was born in 1822 and quickly showed his poly-mathic tendencies, reading medicine and mathematics at the universities of London and Cambridge. Inheriting a fortune from his father, he abandoned his studies and became an explorer, and made important contributions to geographical knowledge of Africa. On his return, he carried out pioneering research in everything from fingerprints to statistics, but his most famous work followed the publication of *The Origin of Species* by his cousin, Charles Darwin. The concept of evolution led Galton to argue for eugenics, in which selective breeding would "improve" the human race. Ironically, Galton never had any children himself. He died in 1911.

3. As the founder of behaviourism, Watson believed even the emotional reactions of humans could be shaped – "conditioned" – by experience. To prove it, in 1920 he performed an experiment in which he succeeded in making an 11-month-old child terrified of a white cuddly rabbit by violently banging a hammer against a steel bar whenever the rabbit appeared.

4. Born in 1905, Harlow's research career began with fairly standard psychological studies of monkeys, devising tests to measure their intelligence, but he found himself intrigued by the monkeys' emotional behaviour. By the late 1950s he began the experiments which made his reputation. Ironically, despite being famed for showing the importance of affection, Harlow was a workaholic who ignored his children.

5. In his 1975 book *Sociobiology*, Wilson attempted to show how the genetic influence on behaviour was a thread connecting social creatures like ants, fish and birds to humans. For example, battles over territory were explained in terms of the genetic concept of kin selection, according to which creatures sacrifice themselves to save more copies of their kin's genes.

Further reading

Nature via Nurture by Matt Ridley (HarperCollins, 2003)

The Blank Slate by Steven Pinker (Penguin, 2002)

A Life of Sir Francis Galton by Nicholas Wright Gillham (Oxford, 2001)

http://www.psy.fsu.edu/history/wnk/ape.html

9
The Selfish Gene

IN A NUTSHELL

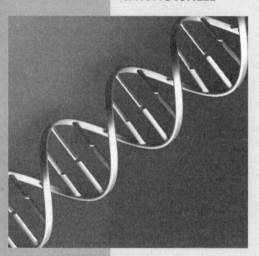

According to the Selfish Gene Theory, Darwin's Theory of Evolution is best thought of not as a struggle for the survival of a species, or even an individual – but as a struggle between genes. While genes are really just strings of chemicals capable of being copied and passed to offspring, the Oxford zoologist Richard Dawkins gave them the metaphorical property of "selfishness". He then showed how their ruthless determination to be passed on could explain apparently senseless altruistic behaviour, whose origins puzzled even Darwin. Not all evolutionary experts accept this "gene-centred" vision of evolution, and the metaphor of the selfish gene has led to much confusion and controversy. In particular, many people think that it implies that we are completely under the control of selfish genes, which undermine our morality and free will. However, as Dawkins himself has tried to make clear, modern humans are in the unique position of being able to rise above the selfish aims of our genes – because we have a mind. It is our mind that allows us to transcend our physical limitations – and it also allows us to go beyond the cold Darwinian morality of genes.

In 1976 a young Oxford University lecturer went public with a disturbing claim about the true nature of humans. "We are survival machines," he declared, "robot vehicles blindly programmed to preserve the selfish molecules known as genes."

This chilling statement appears in the preface to one of the most famous science books of the last 100 years, which propelled its author to international fame. He is Richard Dawkins,

now a professor at Oxford and arguably today's most influential advocate of Darwin's Theory of Evolution.

The book in which he put forward his controversial claim is *The Selfish Gene*, whose snappy title led to huge sales – and even bigger misconceptions. Many believe it is an account of Dawkins' discovery that all living creatures, including humans, are under the control of selfish genes that care only about their own survival, and will do anything to

TIMELINE

1859 Publication of *The Origin of Species* by Charles Darwin, in which he showed how the process of "natural selection" led the emergence of species best suited to their environment.

1871 Darwin publishes *The Descent of Man*, which includes his attempt to solve the problem of altruistic behaviour through the benefit for a group of people, instead of individuals.

1866 Gregor Mendel publishes the results of his experiments, hinting at the existence of genes; they are ignored and forgotten.

1909 Following rediscovery of Mendel's work, Danish geneticist Wilhelm Johannsen coins the term "gene" for the carrier of inherited traits

1953 Francis Crick and James Watson in Cambridge reveal the structure of DNA, and claim this holds the key to the identity of genes.

1962 Vero Wynn-Edwards publishes *Animal Dispersion in Relation to Social Behaviour*, supporting Darwin's view of group selection as cause of altruism.

1964 British theoretical biologist William Hamilton publishes *The Genetical Theory of Social Behaviour*, which explained altruistic traits via "kin selection".

1966 American evolutionary biologist George C. Williams attacks group selection ideas.

1973 Luigi Cavalli-Sforza and Marc Feldman publish first paper on gene-culture theory, which argues that human behaviour can affect genetic traits.

1976 Richard Dawkins publishes *The Selfish Gene*, arguing that evolution works at the level of genes, and their "selfish" determination to be passed on.

1999 Susan Blackmore's book *The Meme Machine* boosts popularity of the concept of memes – cultural "genes" we acquire by imitation.

ensure they are passed on, even sacrificing the "vehicles" they find themselves in – namely, us – if it suits their purpose.

The truth is rather different. *The Selfish Gene* is actually an attempt to show that evolution is best understood by thinking of genes, and imagining them to be "selfish". Dawkins does not claim that genes really are selfish, or that they somehow know what they are doing. He doesn't claim to have discovered the evidence backing this vision of evolution. Least of all does he claim that we humans are powerless to resist the demands of our own "selfish" genes – quite the contrary. Dawkins sought only to bring this vision of evolution before a wider audience, and in so doing coined one of the most potent – but potentially misleading – metaphors in all science.

The selfish gene is an idea that owes its origins to attempts to solve a mystery that had long puzzled evolutionists – including Darwin himself. In his most famous work, *The Origin of Species*, Darwin had shown how the struggle for existence acts as a "filter" through which only some creatures pass – namely, those best able to survive. They would then leave offspring which inherit those attributes, driving evolution onward.

But Darwin knew that some creatures seemed to defy this ruthless logic. They had traits like bravery, which clearly reduced their chances of surviving long enough to have offspring. So why did such traits persist? Why, for example, did birds give out warning cries to others despite the risk of being attacked themselves, or bees indulge in suicidal stinging to defend their nests?

In *The Descent of Man*, published in 1871, Darwin suggested that perhaps the process of evolution sometimes works not at the level of individuals, but on whole groups. So while bravery might not benefit the individual, it could benefit the whole group.

Darwin's idea of "group selection" became very popular among biologists for years afterwards. As recently as the 1960s, the Scottish ecologist Vero Wynn-Edwards was claiming group selection could explain why, for example, rooks stop breeding if their population grows too big. While such behaviour makes no genetic sense for the individual rook, it does makes sense for the survival of the group.

Not everyone was convinced, however: critics of group selection pointed out that perhaps rooks were just incapable of laying lots of eggs when food is scarce. By the mid-1960s, group selection was under attack from a new generation of biologists who insisted that such behaviour – and evolution generally – should be understood in terms of the most basic unit of life: genes.

At the time Darwin was developing his theory, no hard evidence for genes existed (though he knew there had to be some "carrier" for inherited attributes). The first evidence emerged in 1866, in plant experiments by an Austrian monk named Gregor Mendel.

Despite its crucial importance for Darwin's Theory of Evolution, Mendel's work was ignored until long after his death. But by the early 1960s, the structure of DNA had been discovered, and genes were the focus of attention in biology. Among those intrigued by their role in evolution were a small groups of ethologists – scientists studying animal behaviour – who saw genes as the answer to a long-standing mystery: on what, exactly, did "natural selection" operate? The answer seemed to depend on circumstances: sometimes it was individuals, sometimes groups of animals, and sometimes whole species. Yet as critics of Wynn-Edwards had shown, the evidence was rarely convincing.

It was all very messy; how much neater it would be if there was just a single answer. Genes were an obvious candidate for the ultimate agents of evolution, but there was a big problem – the same one that had puzzled Darwin: altruistic behaviour. If genes were the true focus of natural selection, it seemed impossible to explain behaviour that would stop the

JARGON BUSTER

Altruistic behaviour: Acts which do not benefit, or may even harm, individuals, yet which benefit others. Widely seen in animals and humans, such behaviour long posed a serious puzzle for evolutionary biologists.

Group selection: The idea that traits that are harmful to individuals can still spread through a population if they are beneficial to the rest of a group of individuals. Invented to explain how behaviour like altruism can develop, the lack of a clear genetic link means group selection is not thought to be very effective in spreading such traits.

Kin selection: The view that a trait that is harmful to individuals can still be spread through a population if it benefits their genetic relatives. For example, if an altruistic person dies while rescuing three close relatives, one set of genes for altruism has been lost, but three other sets have been saved – a net gain.

Gene–culture theory: The theory that humans benefit from two types of inheritance: genetic and cultural. For example, as well as acquiring genes that led to our ancestors making stone tools, they also "inherited" knowledge of how to use such tools.

Memes: The cultural equivalent of genes, which spread ideas or techniques among minds. Fashions, religions and political movements can all be thought of as based on memes that can be "inherited" by children from the parents.

The strange case of the kamikaze bees

One all-too-familiar example of genetic altruism is the apparent eagerness of honeybees to sting would-be attackers – even though this usually guarantees their own death. Why hasn't evolution "taught" bees that this isn't a great way of passing on their genes? In fact, in most cases it has: out of the many thousands of bee species, fewer than ten have barbed stingers which cause suicidal evisceration. These "worker bees" also share another trait crucial to their bizarre behaviour: they are sterile members of a highly social form of life. Thus suicide does not affect their chances of passing on their genes, as these are zero. But suicidal defence of their hive certainly helps give their fertile relatives a better chance to do their bit and pass on their genes.

genes responsible from being passed on to future generations.

The breakthrough came in 1964, with a paper by a brilliant Oxford theorist named William Hamilton. As a student, Hamilton had read an article by the distinguished geneticist J.B.S. Haldane where he joked that he would lay down his life for two brothers or eight cousins. Behind the joke lay a genetic fact: that brothers all carry copies of their parents' genes. So if one brother sacrifices himself to save the life of his two siblings, only one set of the genes has been lost, and two have been saved – which, from the genes' point of view, is far better than all three being lost. A similar argument applies to cousins, except more lives have to be saved, as they are more distant relatives.

Suddenly the idea of inheriting genes for altruistic behaviour no longer seems so paradoxical. It could make sense if the death of an individual allowed more sets of the same genes to survive. Instead of relying on hand-picked examples from the animal kingdom, Hamilton worked through the consequences of the idea in mathematical detail. He was even able to prove a theorem about how genes for

altruism could exist.[1] This showed that such genes could actually become more common through self-sacrifice – depending on the number of lives saved, how closely related they were, and the cost of the sacrifice.

Hamilton had shown that altruism could be explained genetically via "kin selection", in which creatures sacrifice themselves to save more copies of their kin's genes. It solved Darwin's puzzle, and was later hailed as the greatest advance in evolutionary theory since *The Origin of Species*.[2]

At the time, however, group selection was still influential, and Hamilton had problems getting his ideas accepted. A turning point came in 1966, with the publication of *Adaptation and Natural Selection* by the American evolutionary biologist George C. Williams. This highlighted the shaky evidence for group selection, showing that it is, at best, only a minor effect in evolution. Instead, said Williams, the best way to think of evolution is in terms of selection at the lowest-possible level: genes.

Among those influenced by the work of Hamilton and Williams was Richard Dawkins. Convinced by the elegance and power of the gene-level view, he felt it had to be brought to the notice of a far wider audience. To do that, he had to find ways of bringing the esoteric examples and mathematics to life. He needed a metaphor – and Dawkins hit upon the idea of the "selfish gene".

As a literary device, it was brilliant. It turned abstract genes into undercover agents lurking in our bodies, hell-bent on ensuring their own survival – even at our expense. In his book, Dawkins wielded the metaphor to show how it explained a dazzling range of behaviour in living things, from the shoaling of fish to the bravery of baboons.

But there was one species that Dawkins argued did not live solely by the dictates of the selfish gene: humans. After billions of years of evolution, selfish genes had created a creature with a conscious mind able to defy them.

Attempts to explain human behaviour solely via genes were thus doomed, said Dawkins. Something else was needed, and he argued that it was culture: the collection of beliefs, practices and morals that mould human societies. In this, Dawkins was influenced by the work of, among others, the Italian-born geneticist Luigi Cavalli-Sforza and colleague Marc Feldman, who in the early 1970s developed so-called "gene–culture coevolution". According to this, a cultural shift – such as the advent of diary farming in certain parts of the world – affected the prevalence of genes, such as those needed to digest lactose in milk.

In the final chapter of *The Selfish Gene*, Dawkins suggested an intriguing parallel between genetics and culture: that culture could also be "inherited" via what he called "memes": cultural "genes" passed from parent to child by imitation. Crucially, not all memes would support the aims of selfish genes, however. From people leaving tips at cafes they'll never visit again to children devoting their lives to their sick parents, much human behaviour defies the dictates of the selfish gene. As Dawkins put it in the final sentence of his book: "We, alone on earth, can rebel against the tyranny of the selfish replicators."

Despite this, Dawkins had always worried that his message would be

The Selfish Gene and the Enron scandal

The collapse of the US energy corporation Enron in December 2001 prompted much debate about the corporate culture that prevailed in the company. Enron's chief executive Jeffrey Skilling was a notorious advocate of "Darwinian" management, with constant staff appraisals and the automatic sacking of those with the lowest performance rating. It later emerged that his favourite book was *The Selfish Gene*, which Skilling believed showed that nice guys finished last. Like so many others before him, however, Skilling had been misled by Dawkins's metaphor into believing that selfish genes dictate the behaviour of all living organisms – including humans. Sure enough, it was the "suicidal" action of a whistleblower, Sherron Watkins, that brought Skilling's "Darwinian" company crashing down.

misunderstood – and his fears were soon realised. To many readers, his vivid description of humans as "robot vehicles" that were "blindly programmed" by selfish genes meant we have no way of defying their dictates – precisely the opposite of what he meant. As one reviewer put it: "This is one of the coldest, most inhuman and disorienting views of human beings I have ever heard."

Over the years, *The Selfish Gene* has been attacked on both scientific and moral grounds. Fellow Darwinians criticised Dawkins' apparent insistence that evolution is always best viewed in terms of genes, while one leading philosopher condemned him for propagating a "monstrously irresponsible" view of morality.

While the scientific criticisms do have some justification, most of the invective stems from misconceptions about genes and humans – and Dawkins' failure to shift them. The view that humans must behave as our selfish genes dictate makes no more sense than saying we must walk everywhere because we possess genes for legs. Our minds give us the ability to devise other ways of travelling – and behaving.

Dawkins has spent the last 25 years trying to get across the real message of the selfish gene. Today many if not most evolutionary experts are convinced that Darwin's grand theory is best understood at the level of genes. A growing number are also investigating the implications of Dawkins' fascinating concept of the meme, by which humans "inherit" everything from phobias to religious beliefs.

By coining the metaphor of the selfish gene, Dawkins himself has unleashed a powerful meme on the world. Ironically, it seems to be a meme destined to cause its creator endless frustration.

Notes

1. Like many profound insights in science, Hamilton's mathematical crystallisation of the Selfish Gene concept is remarkably simple. His theorem, often called Hamilton's Rule, states that altruistic behaviour makes genetic sense if the degree of relatedness of the creatures involved, denoted by r, exceeds the ratio of cost, C, to benefit, B. Mathematically, $r > C/B$. For example, imagine two brothers named Adam and Bill. As $r = 0.5$ in the case of brothers, the decision of Adam to lay down his life for Bill – that is, incur a cost of $C = 1$ – only makes genetic sense if the resulting benefit B for Bill is at least doubled.

2. Although rightly credited for his pioneering work, Hamilton was not alone in pondering the mathematics of genetic altruism. In the 1960s a self-taught theoretical biologist named George Price derived more elegant and powerful equations to explain how altruism could thrive in a world apparently dominated by "selfish" genes. Hamilton collaborated with the eccentric American, producing two important research papers. Yet while the insights they contained helped propel Hamilton to the top of the academic profession, their implications had a profound effect on Price, who became an ardent Christian and began giving away all his possessions. In 1974, he finally succumbed to long-standing depression and killed himself in an abandoned building near London's Euston station.

Further reading

The Selfish Gene by Richard Dawkins (Oxford University Press, 1989)

The Darwin Wars by Andrew Brown (Simon & Schuster, 1999)

Introducing Evolution by Dylan Evans and Howard Selina (Icon, 2001)

THE EARTH

10
Catastrophism

IN A NUTSHELL

Legends of the Earth suffering cataclysms such as the Great Flood have circulated for millennia, but attempts to understand them have fallen foul of both the church – which sees them as proof of divine intervention – and the scientific community, which regarded them myths. By the early nineteenth century, the leading geologists of the day such as James Hutton and Charles Lyell had rejected such "catastrophist" views of the history of the Earth. Instead, they argued, our planet has been shaped by slow, steady forces visible today, such as erosion. This "uniformitarian" view of Earth history remained dominant for the next 150 years. Then compelling evidence emerged that the Earth was subject to rare but globally devastating catastrophes which wiped out sizeable fractions of life. Their origin lay not in a vengeful god, but in the cosmos, in the form of impacts from asteroids and comets. One such impact around 65 million years ago is now widely believed to have helped push the dinosaurs into extinction.

Scientists now recognise that a combination of uniformitarianism and catastrophism holds the key to the past, present and future of the Earth, and the life-forms upon it.

From ancient Babylonian manuscripts to Dark Age chronicles and New Age prophecies: all speak of catastrophes striking the Earth out of the blue, with devastating consequences for life on Earth. Many describe global inundations – floods that engulf continents, wiping entire civilisations from the face of the planet. Others describe fires that rain down from the skies.

The Epic of Gilgamesh, written over 4000 years ago, describes how "The Seven Judges of Hell" raised their torches, lighting the land with flame, and then sent a storm that turned day into night. Writing over 2000 years later, the British cleric Gildas described "a fire that fell from heaven" around the year 441 AD that led to dark skies and migrations out of England, with the land still in ruins a century later.

TIMELINE

1694 Astronomer Sir Edmond Halley warns of cata-strophic effect of a comet impact, triggering a flood of biblical proportions.

1795 Scottish geologist James Hutton publishes two volume work laying the foundations of unifor-mitarianism, stressing gradual, steady changes in geology.

1812 French zoologist Georges Cuvier puts forward case for several devastating extinctions of life during history of the Earth.

1830 Charles Lyell's hugely influential work *Principles of Geology*, inspired by Hutton's uniformitarian views, shifts thinking away from catastrophism.

1906 Mining engineer Daniel Barringer publishes evidence that Arizona's famous Meteor Crater really was created by a meteor impact, not volcanic activity.

1942 Harvey Nininger warns of dangers of Earth-impacting asteroids, and claims they may be responsible for strange "breaks" in geological strata.

1956 M. W. Laubenfels at Oregon State University suggests cosmic impact may have triggered extinction of the dinosaurs

1980 Luis Alvarez and colleagues put forward evidence linking extinction of the dinosaurs 65 million years ago to a cosmic impact.

1990 The 180-km wide Chicxulub Crater on the Yucatan Peninsula of Mexico identified as most likely point of impact that pushed dinosaurs into extinction.

1994 Astronomers watch Comet Shoemaker-Levy 9 smash into Jupiter with appalling violence.

2003 NASA's Spaceguard Survey catalogue includes 650 of the estimated 1100 Near-Earth Objects larger than 1 km across.

But these tales of catastrophe have something else in common: until recently they were regarded as tales of divine punishment used by religious leaders to keep their followers in line. Few scientists took them seriously,

regarding them as hang-overs from a superstitious age.

No longer. Today these ancient accounts are regarded as potentially valuable evidence for events that have played a key role in the history of our planet, from its very formation 4500 million years ago. These events are every bit as catastrophic as the legends claimed: cosmic impacts that have dealt severe blows to life many times in the past – and will do so again.

The dramatic view of Earth history now emerging could hardly be more different from that held by many scien-tists even as recently as the 1980s. For almost 200 years, the idea of cosmic events affecting life on Earth was viewed as heretical by the church – which regarded catastrophes as examples of divine intervention – and the scientific establishment, which dismissed them as myths. Those brave enough to put forward evidence for such catastrophes frequently fell foul of both. Yet in the end, the sheer weight of evidence has swept away all doubt about the reality of global catastrophes.

Attempts to make scientific sense of the many legends of global catas-trophes date back to the dawn of modern science itself, in the seven-teenth century. Following the publi-cation of Newton's laws of motion and universal gravitation in 1687, the Oxford University mathematician Edmond Halley decided to apply them to the mystery of comets. By studying records of the appearance of these seemingly capricious denizens of the solar system, Halley argued that the bright comets of 1456, 1531, 1607 and 1682 were in fact one comet, following a vast elliptical orbit around the sun in agreement with Newton's laws. But Halley noted something else as well: a comet crossing the orbit of the Earth might one day collide with us, with devastating consequences.

In December 1694, Halley gave a lecture at the prestigious Royal Society of London in which he claimed that the biblical Great Flood may have been caused by the impact of a comet. He even argued that such huge features as the Caspian Sea might have been formed by such cataclysmic events. Within days, he had recanted his claim – apparently under ecclesiastical pressure. The church viewed catastrophes like the Great Flood as proof of the awesome power of God, and attempts to explain them using science were not to be encouraged. Even so, Halley's lecture was a turning point, highlighting for the first time the vulnerability of the Earth to cosmic forces.

Over the next century, others would also attempt to put biblical accounts of catastrophes on a scientific basis, with the Cambridge mathematician William Whiston claiming that an analysis of cometary orbits showed that the Great Flood occurred in 2349 BC.

By the early nineteenth century, the eminent French zoologist Georges Cuvier appeared to have found rock-solid evidence for the Great Flood – quite literally. By studying the geological strata around Paris, Cuvier had found that fossils of sea creatures in one ancient layer of chalk were over-layed by those of land creatures. Then, just as abruptly, the layer above contained sea creatures again – with the uppermost layer showing evidence of a vast and rapid inundation around present-day Paris. Cuvier took these sudden changes in the fossil record as evidence for sudden catastrophes which devastated life on Earth, of which the Great Flood was just the most recent example.

Cuvier's discoveries, published in 1812, won backing from a number of eminent scientists such as the distinguished geologist Sir James Hall, and the "catastrophist" movement was born. Others were deeply sceptical, however, pointing out that the evidence of a global flood was far from conclusive. Most sceptical of all were the followers of the Scottish geologist James Hutton, founder of the so-called "uniformitarian" view of Earth history. In 1795, he had published a two

JARGON BUSTER

Uniformitarianism: The view of Earth history put forward by the eighteenth century Scottish geologist James Hutton that processes that shape our planet today, such as erosion and vulcanism, were also crucially important in the distant past. Taken up by later geologists, uniformitarianism hardened into the view that global catastrophes were little more than frightening superstitions.

Catastrophism: The view that catastrophic events of biblical proportions, such as the Flood, played a key role in the past history of the Earth. While the roots of catastrophism date back thousands of years, it fell out of favour during the early nineteenth century. Recent recognition of the role of cosmic impacts in Earth history have led to its revival, albeit in far more scientific form.

NEOs: Near-Earth Objects, in the form of comets and asteroids whose orbits around the Sun brings them more than usually close to the Earth (within around 50 million km). Jostled by the gravity of the planets, NEOs can end up on collision courses with the Earth, making them a potential source of global catastrophe.

The Tunguska Impact of 1908

On the morning of 30 June 1908, villagers in a remote part of north-eastern Siberia witnessed the terrifying power of catastrophism at first hand. A huge fireball tore across the sky, accompanied by a deafening roar and huge plumes of black smoke. As it approached the ground, the fireball detonated in an explosion heard over 800 km away. Seismometers detected the reverberations circling the whole planet.

Not until 1927 did scientists finally reach the epicentre of the event, north of the Tunguska River. They found a scene of utter devastation covering hundreds of square kilometres, with the charred remnants of trees lying like matchsticks, all pointing outward from the centre of the blast.

The cause is now thought to be the break-up of a fifty metre wide meteor high in the atmosphere. Despite its small size, the meteor's entry speed of around 100,000 km/hr ensured that when it broke up the resulting shockwave possessed the energy of a 15 megaton H-bomb.

volume text based on the view that the slow, steady processes that shape our planet today, such as erosion, were also crucially important in the distant past.

Uniformitarianism was a powerful idea, allowing geologists to extrapolate from what they could see today to probe questions about the Earth's history. Its most influential advocate was Charles Lyell, a fellow Scottish geologist who saw catastrophism as an attempt by religious zealots to win scientific credibility for their beliefs. Lyell showed that Cuvier's "evidence" for the Great Flood could be explained by gradual changes in sea level, and attacked catastrophists for being too keen on supernatural explanations.

By the 1830s, Lyell's uniformitarianism-based *Principles of Geology* had ensured that claims of global catastrophe were immediately linked to wild-eyed superstition. Yet while Lyell had shown that the evidence for past catastrophes was not conclusive, he had not shown that they were impossible, either. What he had done was to ensure any challenge to uniformitari-

anism would get a very rough ride for the next 150 years.

Hints of where that challenge would come was emerging even as Lyell built his case against catastrophism. For centuries there had been reports of "stones" falling out of the sky, but these had been dismissed as folk stories – until 26 April 1803, when thousands of meteorites rained down on the Normandy village of L'Aigle. An investigation by the leading French astronomer Jean Biot showed that the stones had indeed come from beyond the Earth.

Scientists took much longer to accept that far larger meteors could also strike the Earth – even when the evidence could hardly have been more impressive. In the late 1890s, Grove Gilbert, chief geologist of the US Geological Survey decided to investigate reports of large numbers of meteorites being unearthed around a huge 1.2 km-wide crater in the Arizona desert. Failing to find a giant meteor buried in the crater, Gilbert argued it must have been formed by a bubble of steam in once-molten rock – overlooking the possibility that the meteor could have been vaporised on impact. In 1906, an American mining engineer named Daniel Barringer published evidence linking the Arizona crater to an impact, but failed to convince the scientific establishment (not least because in his scientific paper describing his claim, he derided Gilbert's conclusions). Not until the 1950s would a new generation of geologists recognise the cosmic origins of what is now called "Meteor Crater".

Another clue pointing to a cosmic origin of global catastrophes emerged in the 1930s, with the discovery of the first asteroids on orbits crossing that of the Earth. This prompted the

American astronomer Harvey Nininger to point out that such asteroids could occasionally hit the Earth, with catastrophic results. In 1942, he suggested that such impacts might explain the otherwise puzzling gaps in the fossil record, which hinted at mass extinctions of life several times over the last few hundred million years.

By the mid-1950s, some scientists were suggesting that a meteor impact 65 million years ago might have ended the 100-million year reign of the most successful creatures ever to live on Earth: the dinosaurs. While intriguing, however, the idea remained speculative, and the evidence inconclusive.

That changed in 1980, when a team of scientists from the University of California at Berkeley unveiled evidence powerful enough to break the 150-year stranglehold of uniformitarianism. Led by the Nobel Prize winning American physicist Luis Alvarez, the team had been studying the fossil record of the extinction of the dinosaurs. They found that clay samples from around the time of the event contained very high levels of iridium – an element relatively common in meteors.

The implications of such high levels of iridium were dramatic, pointing to the impact of a huge meteor around 5–10 km across. Such an event would have triggered a global conflagration, with smoke and debris blotting out sunlight for months, causing the collapse of food-chains. Could this explain the disappearance of the dinosaurs?

The mere hint of a return to catastrophism provoked a bitter response from many scientists, with one leading earth scientist at Berkeley dismissing the evidence as "codswallop". Critics argued that the iridium could have come from huge volcanic eruptions

Catastrophism isn't all bad news

Despite its apocalyptic image, catastrophism undoubtedly has its upside. Indeed, we may owe our existence to the disaster that struck the planet 65 million years ago, resulting in the death of the dinosaurs. Their removal gave mammals – which had emerged around the same time as dinosaurs – a chance to thrive. Within ten million years they were the dominant life-forms on the planet.

Recent research suggests even the dinosaurs have reason to be glad of catastrophism. In 2002, an international team of scientists published evidence suggesting that a comet struck the Earth around 200 million years ago, killing more than 50 per cent of all land species. But among the survivors were crocodile-like beasts which seized their chance and evolved to produce the first dinosaurs.

known to have taken place around the same time, and pointed out that the dinosaurs were already dying out before the supposed "impact". Even so, further evidence pointing towards a global catastrophe 65 million years ago started to emerge. In 1988, an international team of scientists revealed the existence of a layer of soot just above the iridium layer at many sites around the world, apparently caused by a global fire. Others found evidence of a huge tsunami created in the Gulf of Mexico around 65 million years ago.

The clincher came in 1990, when American geophysicists revealed the existence of a vast circular structure buried on the coast of Mexico. Measuring 180 km across, the Chicxulub Crater is just the right size, and in the right place, to explain all the other anomalies – and it too is 65 million years old.

Faced with such compelling evidence, most scientists now accept that a huge meteor did strike the Earth 65 million years ago, triggering a global catastrophe – and that it could happen again.[2] While uniformitarian processes like erosion undoubtedly play the major role in shaping our planet, catastrophist events have clearly had a dramatic effect as well. The fossil record shows evidence of at least five mass extinctions of life, with the largest taking place 250 million years ago, eliminating 70 per cent of land animals and 95 per cent of marine life.

The re-emergence of catastrophism as a scientifically credible concept was underlined in spectacular fashion in July 1994. Comet Shoemaker-Levy 9 struck the planet Jupiter with greater violence than the detonation of the entire world's arsenal of nuclear weapons.

Many scientists hoped the event would alert governments to the need to take action to prevent humans suffering the same fate as the dinosaurs. Yet so far, attempts to deal with the threat of cosmic catastrophism have been muted. To date, projects such as NASA's Spaceguard Survey have identified barely half of the estimated 1100 "Near-Earth Objects" (NEOs) whose impact would trigger a global catastrophe. Plans to take action to deal with objects found to be on a collision course are even less well-established.

Almost a quarter of a century after the scientific community shook off the comforting illusion of uniformitarianism, it seems that governments have yet to grasp the terrifying implications.

Notes

1. The idea that a single cosmic punch killed off the dinosaurs is attractively simple, but almost certainly simplistic as well. By the time they finally vanished around 65 million years ago, they had been at the top of the evolutionary tree for 135 million years – during which they are likely to have experienced several global catastrophes, such as climate change, massive volcanic eruptions and cosmic impacts. They may thus have required a "multiple whammy" before succumbing. There is good evidence that at the time of the Chicxulub impact, catastrophic volcanic activity was underway, probably as a result of a "mantle plume" melting through the Earth's crust under what is now India. Some dinosaur experts have also claimed that even ten million years before the impact, the dinosaurs were showing signs of a loss of diversity, suggesting they were gradually dying out in any case. However, this now appears to have been based on over-interpretation of limited evidence. An analysis based on global data published in 2005 by David Fastovsky of the University of Rhode Island showed that by the time of their extinction, dinosaurs were more diverse than ever.

2. While most scientists accept that a meteor impact contributed to the extinction of the dinosaurs, controversy still surrounds the significance of the Chicxulub crater. In 2004, Princeton University geoscientist Gerta Keller claimed to have found evidence that the crater was formed around 300,000 years before the disappearance of the dinosaurs, and that its geological features are hard to square with the standard impact scenario. According to Keller, the evidence bolsters the view that impacts alone cannot have knocked the dinosaurs off their evolutionary perch.

Further reading

The Cosmic Winter by Victor Clube and Bill Napier (Blackwell, 1990)

Life: An Unauthorized Biography by Richard Fortey (Flamingo, 1998)

Catastrophism: Killer Asteroids in the Making of the Natural World by Richard Huggett (Verso, 1997)

11
Plate Tectonics

IN A NUTSHELL

For centuries people had noticed that South America seemed to fit into Africa like two pieces of a jigsaw. Few were willing to see this as anything but a coincidence until the German meteorologist Alfred Wegener showed that fossil and geological evidence pointed to the same conclusion: that all the continents were once part of a huge "super-continent" he named Pangaea, which broke up around 200 million years ago. Unfortunately for Wegener, all his evidence could be explained in terms of other, apparently more plausible, theories. Worse still, his attempts to explain the force propelling the continents were demonstrably incorrect. Not until the 1950s, almost 30 years after his death, did evidence emerge from deep-sea studies to back the idea of continental drift. This included geological studies of the Mid-Atlantic Ridge down the centre of the Atlantic, suggesting that new rock was emerging from deep within the Earth and was pushing America and Europe apart. This led to the concept of plate tectonics, which explains features of the Earth's surface via events at the boundaries of gigantic "plates" floating over the Earth's molten interior. Satellite evidence has now proved that continents really do move at up to 15 cm a year – providing the definitive evidence that forever eluded Wegener.

"Terra firma": our belief in the unchanging nature of the Earth beneath us is proverbial. It is also an illusion, as thousands discover to their horror every year when the Earth beneath them heaves in the throes of an earthquake. Those who survive often talk of the primordial fear they felt on realising that the Earth beneath them is moving.

For all our sophistication, the need to believe we live on a stable planet remains strong. Perhaps this explains why so many came across evidence hinting at the movement of whole continents across the Earth, yet refused to take it seriously. In 1912, a German university lecturer named Alfred Wegener decided those clues could be

TIMELINE

back his ideas was impressive, and resolved many long-standing mysteries. Today it forms the foundation of plate tectonics, a crown jewel of modern science which encompasses everything from the formation of volcanoes and the causes of earthquakes to the origin of mountains. At the time, however, Wegener's claims were dismissed out-of-hand by the scientific community.

It is tempting to blame Wegener's struggles on the blinkered attitudes of academics who resented challenges to their authority. The real reasons are more complex and more interesting, for they highlight the fact that in science, merely being right is not always enough.

Wegener's claims were certainly not new. As long ago as 1596 the Flemish map-maker Abraham Ortelius pointed out how South America seemed to fit into Africa like two pieces of a jigsaw puzzle, and suggested that the Americas had been "torn away from Europe and Africa ... by earthquakes and floods". Others noticed it too, including the Elizabethan scholar Francis Bacon, but if they had ideas about its origins, they kept them to themselves. Not until the mid-nineteenth century did anyone pluck up the courage to take the coincidence seriously. In 1858, the French geographer Antonio Snider-Pellegrini pointed out another "coincidence": fossils of the same plants appeared on both sides of the Atlantic; so did geological formations such as coal deposits. Snider-Pellegrini concluded that America, Africa and Europe had once been joined together in a vast continent, and claimed that the break-up of this continent could explain the biblical flood. At a time when geolo-

ignored no longer, and put forward hard evidence that the continents of the Earth are moving across our planet like gigantic rafts of rock.

According to Wegener, they had once all been part of a giant supercontinent that broke up hundreds of millions of years ago, and have been on the move ever since, their collisions creating vast mountain chains.

Put so bluntly, Wegener's claims sound like the fantasy of some wild-eyed crank. Yet the evidence he had to gists were turning against such "cata-strophic" accounts of Earth history, this was a bold claim – and one all but guaranteed to achieve immediate obscurity.

Yet the basic idea of drifting conti-nents refused to die. In 1908 the American geologist Frank Taylor pointed out that collisions between drifting continents could explain the origin of mountain belts such as the Himalayas and Alps. To most geolo-gists, however, this was the answer to a problem already resolved by the then-popular view that the Earth was shrinking. According to this, the once molten Earth is still cooling down, and shrinking in size. As a result, the Earth's surface is being squeezed into an ever smaller area, causing wrinkling – or, as we call them, mountain chains.

It was against this theory that Alfred Wegener, a 32-year-old lecturer in meteorology at the University of Marburg, Germany, pitted himself in resurrecting the idea of continental drift. He had first noticed the jigsaw-like fit of the continents in 1903 while still a student, but dismissed it as a coincidence. That changed in 1911, when Wegener read of the oddly similar fossils on either side of the Atlantic that had intrigued Snider-Pellegrini.

The conventional explanation was that these were the result of vast "land-bridges" that had spanned the ocean millions of years ago, just as Central America now links North and South America today. The similarities in the fossil and geological record extended to more than just Africa and America, however. For example, a fossil fern named *Glossopteris* had been found on both continents – and also in India, and Australia. Then there was the evidence for simultaneous ice ages in all these regions around 290 million years ago – again, hard to understand unless they were all once close together. Wegener decided that the simplest explanation was that all the continents had once formed part of a super-continent he named Pangaea (from the

JARGON BUSTER

Lithosphere: uppermost layer of the Earth, including the crust and the rigid top-most layer of the mantle. Broken up into nine major "plates", plus fourteen or so smaller ones.

Asthenosphere: the 300-km deep "sea" of partially molten mantle rock on which the litho-sphere floats. It begins around 100 km below the surface.

Pangaea: a vast super-continent formed around 320 million years ago from the collision of two huge areas of continental rock. Around 200 million years ago, Pangaea split into two again, with Laurasia in the north eventually forming America and Europe, while Gondwana in the south formed Africa, South America, India and Antarctica.

Subduction zone: Region where one plate slips beneath another, creating an oceanic trench.

Mid-oceanic ridge: region where two conti-nental plates separate, allowing fresh basaltic rock to rise to the surface from deep within the Earth. The result is the longest chain of moun-tains on the Earth, stretching for 40,000 km, with the highest peaks.

Geomagnetic reversals: The sudden flip-over of the Earth's magnetic field, with the north and south magnetic poles swapping places. It leaves its effects in molten rock, where it played a key role in confirming continental drift. The most recent reversal took place 250,000 years ago, but despite being recognised for almost a century, its precise cause remains unclear.

The nuclear reactor beneath our feet

Miners have long known that it gets hotter the deeper into the Earth they dig, but the source of that heat was identified barely a century ago: radioactivity. Natural uranium, thorium and potassium isotopes trapped inside the Earth at its formation have turned our planet into a seething nuclear reactor. The crust acts as a "containment vessel", albeit a pretty ineffective one: each year tens of thousand of people die from the radioactive "pollution" seeping out of the Earth in the form of cancer-causing radon gas.

Greek for "All Land") that split up around 200 million years ago.

In January 1912, Wegener gave his first lecture on his theory of continental drift; like his predecessors, he found little interest. In 1915, he brought out a small book entitled *The Origin of Continents and Oceans*, with which he hoped to attract the interest of the Earth sciences community. Published in German, it too failed to ignite much interest. To professional geologists, Wegener was an outsider claiming to solve problems they had already solved, but using an idea that raised at least as many new questions. For how could continents simply drift around the world? How were they being propelled ?

Wegener was well aware of the problems facing his theory, and over the next decade he issued revised accounts addressing them. Chief among them was the need to explain the forces propelling continental drift. Wegener argued that the continents were vast areas of a granite-like material called sial, which floated on a denser but softer material called sima, which made up the ocean floor.

To explain the driving force of continental drift, Wegener drew on his meteorological knowledge to suggest that they were due to the rotation of the Earth – a key driving-force of weather systems. The so-called Eötvös Force was known to be capable of driving air westward and away from the poles. Wegener claimed that the same "pole-fleeing" force might propel the continents, along with a kind of tidal effect similar to that which affected the oceans.

Yet, far from impressing the geophysics community, Wegener's attempts to explain continental drift simply gave them clearer targets to aim at. Studies of the ocean floor proved that it was not remotely soft like sima, while detailed calculations showed his proposed forces were too feeble to propel the continents. In contrast, the Shrinking Earth theory had no problems generating sufficient force for mountain-building.

By 1929, as he finished the fourth edition of his book, Wegener conceded that he had still not come up with a rock-solid case. It was to be his final statement on the subject. The following year, he set off on an expedition to Greenland; he never returned. By the time of his death, continental drift was as far from being accepted as ever, with academics dismissing it as everything from "very dangerous" to "utter, damned rot".

Wegener's book had attracted some supporters, however, including the distinguished British geologist Arthur Holmes, who in December 1927 put forward the first glimmerings of what was to become the true explanation of continental drift. Holmes was an expert in radioactivity, then emerging as a key issue phenomenon in geophysics. Scientists had found that uranium and other elements trapped in the Earth were keeping its interior hot. This was bad news for the Shrinking Earth Theory, as it meant that the Earth was neither cooling nor shrinking. In contrast, it was potentially good news for continental drift, as the heat might generate vast circulating loops of molten rock inside the Earth, capable of moving whole continents.

It was still too much for most geologists, however, and with Wegener's death his theory slipped once more into academic obscurity. Its triumphant revival followed discoveries made about the one part of the Earth neither Wegener nor any of his contemporaries really understood: the ocean floor.

During the 1950s, studies of the Mid-Atlantic Ridge, the vast range of submerged mountains running down the centre of the ocean, revealed V-like structures that looked suspiciously like a rift between two vast plates.[1] Studies of the sea bed also revealed that it was much younger than anyone thought: just 200 million years or so – curiously similar to the date at which Wegener's "super-continent" supposedly split apart.

In 1960, the American geologist Harry Hess at Princeton University connected these clues with Holmes's old idea about convection inside the Earth. The result was the theory of "seafloor spreading", in which new crust emerged at the Mid-Atlantic Ridge, riding on loops of molten rock, and then sank again at trenches. In contrast to Wegener's view of continents sliding over the ocean floor, Hess's theory held that they were being pushed apart by new material spewed upwards at oceanic ridges.

Support for Hess's view quickly emerged. Measurements of the magnetic field on the floor of the Pacific revealed a bizarre stripe-like pattern of alternating north and south

The mystery of the Earth's magnetic flips

Despite their value in the debate over plate tectonics, the reversals in the Earth's magnetic field are still somewhat mysterious – not least because of their erratic timing. The last flip took place around 775,000 years ago, but the intervals between flips can be anything from 40,000 to a million years. Using supercomputer simulations of conditions inside the Earth, geophysicists have found evidence that the flips are the result of the complex interactions between the magnetic fields in the liquid outer core and solid inner core. The latter acts as a stabilising influence, but every so often it vanishes – allowing the outer core field to flip.

poles. Geologists already knew that the Earth's magnetic poles occasionally flip over, and that hot rock retains a record of the Earth's magnetic field as it cools. In 1963, Drummond Matthews and his student Fred Vine at Cambridge University pointed out that the stripe pattern was consistent with fresh, hot rock continually emerging from within the Earth over 200 million years – just as Hess's theory required.

By the late 1960s, the evidence for continental drift was becoming hard to dismiss. Geologists began to take seriously the implications of "plate tectonics", the processes that take place where two plates meet. They found that the theory allowed them to make sense of a plethora of features of our planet. A titanic collision between India and the Asian plate around 45 million years ago explained the origin of the Himalayas, while the plethora of volcanoes around the Pacific Rim – the so-called "Ring of Fire" – bears witness to the destructive forces at work at the edge of the Pacific plate. High risk earthquake zones like the San Andreas Fault of California proved to be regions where two plates slide against each other.

Today the evidence for continental drift is beyond all doubt. The Earth's surface is now known to comprise of 10 major plates and fourteen smaller ones, ranging from 50 km thick beneath the oceans to up to 250 km thick below continents. By bouncing laser beams off orbiting satellites, scientists have also proved that the plates are moving, at up to 15 cm a year; North America and Europe are currently separating at around 2 cm a year.[2]

Even today, however, some mysteries remain. What created the pattern of plates now covering the Earth? What triggered the break up of Pangaea? Even the nature of the forces propelling the plates – the problem which stymied Wegener – is not entirely clear. Current thinking suggests that the sinking of the old plates is the principal driving-force – but the jury is still out.

What is clear is that Wegener's critics were not merely reactionary die-hards. Their criticisms were largely perfectly reasonable, and Wegener's responses often unsatisfactory. Ultimately, his greatest failing was one he could do nothing about: he was simply too far ahead of his time.

Notes

1. Exactly when plate tectonics started on Earth is still a hotly debated question among geophysicists. Ancient sea floor material bearing the signs of tectonic processing has been dated at around 1.9 billion years old. In 2000, Timothy Kusky of St. Louis University identified rock near the Great Wall of China that appears to be more than 500 million years older still. If so, this would place the origins of plate tectonics to no later than two billion years after the formation of the Earth.

2. Plate tectonics are still modifying the appearance of our planet. Calculations by Professor Christopher Scotese and his colleagues at the University of Texas suggest that around 50 million years from now, the Mediterranean Sea will have vanished, Africa having smashed into the landmass of Europe and closed up the sea-filled gap in between. Around 250 million years hence, the Americas and Eurasia will have completed a pincer movement trapping Africa between them, forming a huge "supercontinent" which Scotese has called Pangaea Ultima (see www.scotese.com).

Further reading

The Earth: An Intimate History by Richard Fortey (Perennial, 2005)

The Dating Game: One Man's Search for the Age of the Earth by Cherry Lewis (Cambridge University Press, 2000)

Earth System History by Steven M. Stanley, (W.H. Freeman, 1999)

MATHEMATICS THAT COUNTS

12
Bayes's Theorem

IN A NUTSHELL

If we want to test a belief or theory, we need evidence – yet not all evidence is equally convincing. For example, if a drug cures 80 per cent of patients, that's clearly more impressive evidence than if it only cures 60 per cent – but just how much more convincing? The answer comes from a mathematical method devised over 200 years ago by Thomas Bayes, an English clergyman and mathematician. Known as Bayes's Theorem, it shows how to update our beliefs in the light of new evidence. To do this, we must first state our current level of belief, based on what we already know. Yet in the absence of any previous research, this "prior belief" may be little more than a guess. This led to Bayes's Theorem being branded as hopelessly subjective and unscientific, and until the 1980s it was all but abandoned by scientists. Statisticians have since found ways around the problems of prior evidence, and Bayes's Theorem is now increasingly recognised as the most reliable way of extracting insights from a morass of complex data.

Medical researchers announce a breakthrough in treating heart attacks, claiming it halves death rates. A forensic scientist tells a jury that DNA evidence points to odds of millions to one that the accused is guilty. The weather forecast predicts a beautiful weekend.

Every day, we are bombarded with claims and counterclaims, all apparently based on the latest scientific evidence. Clearly, many of them are just plain wrong – but which ones? Some seem very plausible, while others fly in the face of past experience. It's rare that new medical therapies halve death rates, so any such claims need to be viewed with suspicion. On the other hand, weather forecasts are getting more reliable – so perhaps we can plan that weekend break.

But how can we take account of something as vague as plausibility in our decision making? Should we even do so? Should we allow it to sway our decisions? In a society increasingly forced to make judgements based on scientific evidence, such questions have never been more important.

Astonishingly, the key to answering them has been known for over 200

TIMELINE

1763   First appearance of Bayes's Theorem, in a paper published two years after the death of its author, the Rev. Thomas Bayes.

1774   French mathematician Pierre-Simon Laplace puts Bayes's Theorem in its modern form.

1922   British mathematician Ronald Fisher introduces supposedly objective methods of testing scientific theories, known as "significance testing".

1926   Cambridge mathematician Frank Ramsey publishes Dutch Book theorem, showing rational beliefs follow laws of probability.

1928   Romanian statistician Jerzy Neyman and British statistician Egon Pearson put forward "frequentist" methods of hypothesis testing.

1939   Criticism of Fisher-Neyman-Pearson methods made by leading statisticians Harold Jeffreys in Cambridge and Carrodo Gini in Italy.

1940   At Bletchley Park, Alan Turing uses Bayesian methods to break Nazi Enigma codes.

1950s   Bayesian methods fall out of favour with scientists, as journal editors demand use of "objective" significance and hypothesis tests.

1960s   Warnings about dangers of frequentist methods re-emerge.

1980s   Increasing computer power starts to revive Bayesian methods; criticisms of frequentist methods become more vocal.

1992   Fears over interpretation of DNA evidence lead to use of Bayes's Theorem in jury trials.

1990s–present   Huge revival of Bayesian methods, now widely used in areas ranging from medical research to business sales analysis.

evidence. From scientists to jury members, code-breakers to consumers, everyone can benefit from its power.

Its origins belie its extraordinary potential. In 1763, the prestigious Royal Society published a 49-page paper entitled "Essay Towards Solving A Problem in the Doctrine of Chances". Its author was Thomas Bayes, an English clergyman who had died two years earlier. The paper set out an attempt to solve a key issue in science: working out the impact of new evidence on the chances of a theory being correct.

For example, imagine you suspect that a football referee is using a coin biased towards heads. To find out, you carry out 100 tosses, and get fifty-nine heads. If the coin was fair, you'd expect around fifty, so clearly there's some evidence supporting your theory; the question is, how impressive is it?

In Bayes's day, mathematicians could only answer a less interesting question: what are the chances of getting so many heads, *assuming* the coin is completely fair? But that's not the issue: we want to know whether getting fifty-nine heads means the coin *really is* fair.

Bayes showed that it was possible to answer such questions using a technique now known as Bayes's Theorem. Put simply, it shows that the odds of a theory being true in the light of evidence depend on two factors. The first measures the "strength" of the evidence, and compares the relative likelihood of getting such evidence if the theory is true and if it were false. As one expects, the larger this so-called likelihood ratio, the more impressive the strength of evidence.

There is a second factor, however – and one which has made Bayes's Theorem very controversial. This is the

years, but for decades has been mired in controversy. It is still regarded by some as scientifically heretical, even dangerous. Yet on paper, it sounds innocuous enough: a mathematical recipe for turning evidence into insight.

It is known as Bayes's Theorem, and after years in the scientific wilderness, it is now being recognised as the most reliable way of making sense of so-called prior probability; that is, the probability of the theory being true, prior to the data being collected. On the face of it, this sounds bizarre: as the whole point of collecting data is to find out if the theory is true, on what evidence should this "prior" probability be based?

Bayes himself tried to tackle this problem by arguing that in the absence of any evidence, the prior probability could be anywhere between zero and one. He couldn't solve the resulting mathematics, however – which may be why his results weren't published until after his death in in 1761.

A decade later, the great French mathematician Pierre-Simon Laplace took on the problem, solved it and cast Bayes's Theorem in its modern form. He then applied it to the real-life problem of investigating claims that significantly more boys were being born in Paris than girls. Laplace used birth statistics and Bayes's Theorem to confirm the suspicions.

Backed by the authority of Laplace, Bayes's Theorem became the standard way of assessing scientific evidence until the early twentieth century when a number of leading statisticians began to question "Bayesian methods". Their misgivings centred on the old issue of setting prior probabilities – which seemed to allow anyone to reach any conclusion from the same data. For example, sceptics of a new medical treatment would give it a prior probability of success far below that given by its supporters – instantly making apparently convincing evidence merely suggestive.

This smacked of scientific anarchy, and in the 1920s some statisticians tried to develop entirely objective ways of testing theories which eliminated the need for prior probabilities. In reality, these techniques just swept the issue under the carpet, creating an illusion of objectivity. Yet despite attempts by some leading statisticians to point this out, the new "frequentist" methods soon became the standard way of assessing scientific evidence.

From the 1930s onwards, "Bayesian" methods fell into disuse, branded as

JARGON BUSTER

Prior probability: The chances of a theory or belief being true, prior to taking into account new evidence. For example, statistics show that the average woman faces a 1 in 100 chance of developing breast cancer during her lifetime. So her prior probability of having the disease before any tests are made is 1 in 100.

Posterior probability: The updated chances of a theory or belief being true, *after* taking into account new evidence – from, for example, a breast screening.

Likelihood ratio: A measure of the strength of evidence supporting a theory or belief. The stronger the evidence, the higher the likelihood ratio: for example, DNA evidence can produce likelihood ratios of 100,000 or more.

Frequentism: The view of probabilities as the frequency of events in the long run. For example, a coin said to have a 50 per cent probability of "heads" will tend to give 50 per cent heads after many throws.

Bayesianism: The view of probabilities as degrees of belief based on all available evidence. According to this, statements about the chances of getting "heads" from a coin must take into account any evidence of effects that might affect them, such as tampering.

The Rev Bayes's magic formula

Bayes's Theorem can be stated in various ways, but the simplest is the so-called "odds form", in which the level of belief in a particular theory is captured by the odds of it being true. In this case, Bayes's Theorem states that when new evidence, E, comes to light, the original ("prior") odds of the theory being correct – Odds(T) – are changed to a new value of Odds(T given E), according to the formula

Odds(T given E) = Odds(T) × LR

where LR is the so-called Likelihood Ratio, given by the probability of E emerging if T is true, divided by the probability of E emerging if T is *not* true. So, for example, if the evidence E is as just as likely to emerge when T is true as when it is false, the LR is 1, and this new evidence has in fact made no difference to the strength of the case for the theory T.

hopelessly subjective and unscientific. A handful of diehards kept the flame alight, however. They included Alan Turing, the brilliant Cambridge mathematician who led the effort to break the Nazi Enigma codes. At Bletchley Park, Turing and his colleague Jack Good used Bayesian methods to work out the most probable German text corresponding to intercepted code. While scientists were decrying its use in their research, Bayes's Theorem was secretly helping to win the war.

The post-war boom in scientific research saw frequentist methods become all but mandatory. Claims for new theories were only taken seriously by other scientists if the evidence achieved "statistical significance". To do this, the data had to be plugged into frequentist formulas and checked to see that they gave a so-called P-value below 1 in 20.

Many scientists thought this P-value was the probability of their results being nothing more than a fluke – so the lower the value, the more impressive the evidence. What they didn't know was that frequentist statis-

ticians, desperate to avoid using Bayes's Theorem, had tried to find an objective way of testing theories but had failed. Instead, they came up with the P-value idea, which was no better than the old formulas giving the odds of getting the data, on the *assumption* they are really just a fluke. Scientists, in contrast, thought the P-value was an objective test of whether a theory was right or not.

It wasn't – and from the 1960s onwards, top statisticians tried to warn scientists of the dangers of misunderstanding P-values, but without success. For most scientists, P-values seemed objective and easy to use and, most important of all, top journals demanded P-values as evidence that the findings were worth publishing.[1]

For years, Bayesian methods were all but ignored by scientists. But during the 1980s developments took place that sparked a huge revival in their use. First, the growing availability of computers meant that anyone could carry out the sometimes complex calculations needed to use Bayes's methods. Second, everyone from scientists to supermarket managers were increasingly being deluged by data, and needed ways of turning it into insights about, for example, which drugs worked, or which products sold. Bayes's Theorem gave them the power to turn mere hunches into hard conclusions – and allowed them to update their conclusions constantly as new evidence came in, which frequentist methods struggled to do.

For example, in 1992 medical researchers in Scotland hit the headlines with a treatment based on a "clot-busting" drug that cut deaths among heart attack victims by around 50 per cent. At least, that is what the frequentist methods seemed to show. Many experts were sceptical of such an impressive success rate. It flew in the face of much bitter experience, yet there seemed no way to take this lack of plausibility into account.

Shortly after the research was published, two medical statisticians, David Spiegelhalter of the Medical Research Council and Stuart Pocock of the London School of Hygiene, used Bayes's Theorem to combine the Scottish findings with prior evidence based on previous research. They concluded that the real effect was likely to be only half as good as claimed. So who was right? Eight years later, in 2000, a team of US researchers reviewed all the evidence from many more studies of the same clot-busting treatment, and concluded that it was indeed only about half as good as the first study had claimed. The supposedly "dangerous" idea of using prior evidence had allowed Bayes's Theorem to alert doctors to the fact that the treatment wasn't the miracle cure it seemed.[2]

During the 1990s, Bayes's Theorem started to make headlines in another field: the legal profession. With the power to combine different sources of evidence, it had obvious uses in jury trials, and lawyers began to use experts in Bayes's Theorem as defence witnesses.

Most of the cases centred on DNA evidence, where suspects were accused of serious crimes based on similarities between their DNA and traces at the crime scene. By the late 1980s, juries were routinely hearing that the odds against getting such similar DNA from someone picked at random were millions to one. But some forensic scientists went further, telling juries that these tiny odds were also the

The illusion of objectivity

Bayesian methods are routinely criticised for introducing a "subjective" element into the process of extracting insight from evidence. Critics argue that in the absence of any previous research, the so-called "prior odds" required by Bayes's Theorem may end up being based on little more than intuition and educated guesswork. Advocates of Bayesian methods point out, however, that this is precisely what happens in real life in any case; Bayes's Theorem simply makes this otherwise vague process both explicit and quantitative. Moreover, while many scientists may dislike the presence of subjectivity in what they do, that does not alter the fact that it is actually *ineluctable*. In 1926 the Cambridge mathematician Frank Ramsey published the so-called "Dutch Book Theorem", which shows that rational assessment of evidence inevitably involves the use of Bayes's Theorem and prior beliefs – which can often be entirely subjective.

chances of the accused being innocent. Bayes's Theorem shows this isn't true: only when all other evidence has been taken into account can the odds against innocence be estimated – and defence lawyers showed that if there is little other evidence, not even DNA matches are convincing enough to convict. Since the mid-1990s, Bayes's Theorem has played a key role in a number of successful appeals against convictions.

Today there are clear signs that Bayes's Theorem is moving back to the centre stage of science. Over the last decade, the number of scientific papers using Bayesian methods has increased ten-fold, with researchers using them to test theories in fields ranging from physics to pharmacology.

The power of Bayes's Theorem is being recognised far beyond academia, however. Major corporations now rely on Bayes's Theorem to extract insights from their own sales and marketing data and allowing them, for example, to target mail-shots far more accurately, producing much less junk mail. Many multinationals use Bayesian-inspired data management software created by Autonomy, one of the most successful hi-tech companies of the late 1990s whose founder, Dr Michael Lynch, became Britain's first dollar billionaire.

After decades in the doldrums, it seems that the power of the Rev. Bayes's mathematical revelation is finally being recognised.

Notes

1. P-values have also helped a lot of total nonsense seep into the scientific literature, through their tendency to exaggerate the "significance" of meaningless fluke results. The use of P-values has proved especially valuable to those seeking scientific backing for such flaky notions as the existence of bio-rhythms and the effectiveness of wishing for good weather. One of the most bizarre examples centres on a study published in 2001 by Leonard Leibovici of the Rabin Medical Centre, Israel, purporting to show the effectiveness of "retroactive prayer". Some early research has hinted that patients may benefit from being prayed for. According to Prof Leibovici's study, published in the *British Medical Journal*, prayers even helped patients who had already recovered. The findings, whose supposed significance was demonstrated using P-values, sparked calls for a complete overhaul in notions of space and time. To statisticians, however, the results are just further proof of the dangers of misunderstanding P-values.

    Concern about use of P-values to back implausible claims is mounting. In 2004, a team from the US National Cancer Institute, Bethesda, Maryland, warned: "Too many reports of associations between genetic variants and common cancer sites and other complex diseases are false positives." It added: "A major reason for this unfortunate situation is the strategy of declaring statistical significance based on a P-value alone."

    Later the same year, the editors of the journal *Nature Medicine* and its sister publications – the most prestigious science journals in the world – announced a crackdown on sloppy statistical techniques. So far, however, they show no signs of being able to give up their addiction to P-values.

2. Hardly a week goes by without some new medical study making headlines with some supposed link between health and, say, mobile phone use or coffee drinking. Some of these claims are worth taking seriously, but many are not; the problem, of course, is telling them apart. Bayes's Theorem can be used to assess the inherent credibility of a medical claim, by taking into account both the strength of evidence of the new findings, and prior insight into such a link. Such "credibility analysis" can help identify studies worth taking seriously, and those likely to prove illusory. An online calculator for assessing medical study results can be found at http://tinyurl.com/6ubfo

Further reading

Scientific Reasoning: The Bayesian Approach by Colin Howson and Peter Urbach (Open Court, 1993)

"Facts and Factions: The Use and Abuse of Subjectivity in Scientific Research" by Robert Matthews, in *Rethinking Risk and the Precautionary Principle* edited by Julian Morris (Oxford 2000), available online at: www.tinyurl.com/4atbd

"Why should clinicians care about Bayesian methods?" by Robert Matthews, in *Journal of Statistical Planning & Inference* 94 pp 43–58 (2001), available online at tinyurl.com/aympu

13
Chaos

IN A NUTSHELL

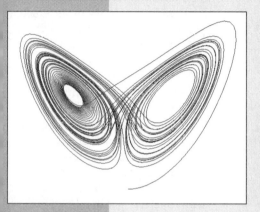

The world around us seems to feature two types of phenomena: the regular and the random. Yet since the 1960s, scientists have become aware of a third and extremely important type: chaotic phenomena, whose apparently random behaviour is actually just extremely complex – and can be forecast, at least to some extent.

The classic example is the weather, which is neither random nor regular, and can be predicted some time into the future. Just how far into the future depends on the strength of chaos present in the atmosphere. Estimates based on computer models of the extremely complex factors involved in creating our weather suggest forecasts will never be reliable beyond around 20 days or so ahead.

The reason is the so-called Butterfly Effect, a hallmark of chaos in which just a tiny change can have major consequences. In the case of weather forecasts, meteorological data is never totally accurate, and the Butterfly Effect amplifies these errors over time until they totally ruin the forecast.

Often appearing wherever many different factors interact, chaos has been found in many natural phenomena, from the orbits of the planets to electrical activity in the brain. Scientists have found that they can sometimes switch chaos on and off – but sadly, no-one has any idea how to do this for the weather.

You're about to set off for the airport when the phone rings. It's a friend who left his watch at your place last night. You fail to find it, and race out of the door – five minutes late. The traffic lights are all against you, and you arrive at the rail station only to see your train leave – and there isn't another for half an hour. By the time you reach the airport, it's too late: the plane has left, and there isn't another until tomorrow. That five minute delay in getting out the door has ballooned into a whole day lost.

It's an everyday example of a phenomenon that has prompted a revolution in science: chaos.

TIMELINE

1961   Ed Lorenz at the Massachusetts Institute of Technology discovers the first evidence for chaos in his computer model of atmospheric convection.

1972   Lorenz coins the term Butterfly Effect to capture the idea of small effects having huge consequences.

1975   American mathematician James Yorke coins the name "chaos" for the behaviour that causes the Butterfly Effect.

1976   Australian physicist Robert May publishes ground-breaking paper in *Nature* alerting scientists to the chaos lurking in even simple equations.

1984   American chaos expert Jack Wisdom and colleagues predict chaotic tumbling in Hyperion, a moon of Saturn.

1987   Observations of Hyperion confirm it as the first example of a chaotically tumbling object in the solar system.

1987   *Chaos: making a new science* by American journalist James Gleick becomes a best-seller, and brings chaos to public attention.

1989   Jacques Laskar of the Bureau des Longitudes in Paris shows that the Earth's orbit around the sun is chaotic, and its position becomes completely unpredictable after around 100 million years.

1990   William Ditto and colleagues at Georgia Institute of Technology begin pioneering studies into controlling chaos.

1990– present   Economists and financial analysts investigate ways of detecting chaos in data, to allow better forecasting.

1995– present   Medical researchers find evidence for chaos in the behaviour of healthy tissue and organs, including the heart and brain.

From the weather to astronomy to medicine, chaos has been found lurking in fields once thought to be well understood. By confounding the notion that small effects can only have small consequences, chaos is now recognised as one of the most important phenomena in science. It sets fundamental limits on what we can know, and can make a mockery of our attempts to predict the future.

Given its ubiquity, the wonder is that no one noticed the existence of chaos sooner. Only in the 1960s did scientists start to discover just how common chaotic behaviour really is. This realisation followed the growing use of computers by scientists, which gave the power needed to probe the kinds of phenomena in which chaos rears its head.

These can sometimes be relatively simple: for example, a pendulum swinging from the end of another pendulum can show chaotic behaviour, with just a tiny change in the starting position producing radically different behaviour. Typically, however, chaos emerges when there are three or more factors interacting with each other in complex ways.

That makes pen-and-paper study of their behaviour impossible; when studied by computer, however, they often spring big surprises – as the American meteorologist Ed Lorenz found in 1961, after the most important coffee break in scientific history.

At the time Lorenz was a researcher at the Massachusetts Institute of Technology, studying simple models of the Earth's atmosphere. For decades meteorologists had dreamed of predicting the weather, but had been stymied by the complexity of the equations governing the atmosphere. These involve quantities like temperature and wind speed linked together in complex ways. As there is no simple straight-line relationship between the quantities involved, mathematicians called them non-linear equations: a 10 per cent increase in, say, temperature doesn't necessarily change wind-speeds by the same amount.

In the winter of 1961, Lorenz was using a computer to tackle weather prediction. While it still could not find the general solution to non-linear equations, the computer could at least allow Lorenz to watch how they behave in specific cases.

Lorenz was especially interested in atmospheric convection, a key meteorological phenomenon. After programming in the equations describing convection, he let his computer print out a graph of what happened over time. As it could only manage 60 multiplications a second, Lorenz decided to speed things up – with revolutionary consequences. Instead of starting each new "run" from scratch, he fed the computer with figures taken from the middle of the previous run, and set it off again. He then went off for a coffee.

When he returned, there was a big surprise waiting for him. Lorenz expected the computer simply to duplicate the second half of the previous run, before chugging off on the next. To his astonishment, the computer refused to repeat what it had done before. It started the same – but then drifted off.

Lorenz's first instinct was to blame a faulty valve. But then he realised he hadn't started off the computer using precisely the same values as it found in the previous run: he'd rounded them up very slightly. He never thought so slight a change would make any real difference – until the computer started to spit out totally different results.

Lorenz had stumbled into a major discovery: with non-linear phenomena like the weather, tiny changes can have huge consequences. The implications were stunning. First, it meant that – in principle at least – the flap of a butterfly's wing in Brazil could trigger a tornado in Texas, an effect Lorenz later immortalised in the term the Butterfly Effect. Second, it cast doubt on the prospects of long-term weather prediction: meteorological data is

JARGON BUSTER

Non-linearity: The property of some phenomena – like the weather, or commuting to work – where small changes don't always have small consequences. For example, an error of just a few per cent in weather data can totally wreck a long-term forecast, while a 10-minute lie-in can lead you to miss your bus – and make you an hour late.

Lyapunov timescale: A measure of the strength of chaos, and thus how long before forecasts become useless. The longer it is, the longer predictions will remain reliable. For example, chaos due to the gravitational pull of the other planets means the Earth's position becomes unpredictable more than 100 million years ahead.

Strange attractor: Roughly speaking, the collection of possible states that a chaotic phenomenon can exist in. In the case of the weather, the strange attractor can be thought of as the climate, which ensures that winters are typically colder than summers, though each day can be radically different. But just a tiny change, such as – in theory at least – the flap of a butterfly's wing, can nudge the weather into another part of the strange attractor, making long-range forecasting impossible.

Butterfly Effect: Picturesque term for what mathematicians call "sensitive dependence on initial conditions": the way in which non-linear phenomena like the weather can be affected by even the tiniest effects – such as the flapping of a butterfly's wing.

Hyperion – the chaotic potato

Orbiting the planet Saturn every twenty-one days, Hyperion was long thought to be a perfectly ordinary moon of the ringed planet until it was photographed by the Voyager 2 probe in 1981. The images revealed Hyperion to be a potato-shaped object which seemed to be rotating in a curious way. In 1984, a team led by theorist Jack Wisdom at the California Institute of Technology suggested that the cause lay in Hyperion's odd shape, plus the gravitational pull of the neighbouring moon Titan, which together were causing the moon to tumble chaotically. Studies of the changes in brightness of Hyperion have since convinced most astronomers that Hyperion really is in the grip of chaos.

never perfect, and the Butterfly Effect meant that even tiny errors can grow with time to ruin a forecast.

Scientists now know that the same applies to a host of phenomena where non-linear effects are at work. Even the positions of the planets – long regarded as the classic example of predictability – have been found to suffer from the Butterfly Effect. All of them are exquisitely sensitive to slight changes or small errors, and eventually become unpredictable, their behaviour becoming apparently completely random. Crucially, however, it isn't true randomness: it's simply that the laws governing such phenomena produce incredible complexity that can look random. The result is what the American mathematician James Yorke in 1975 dubbed *chaos*: a kind of "half-way house" between complete regularity and utter randomness.

Deciding when randomness is actually chaos is clearly crucial, and mathematicians have developed ways of both revealing the presence of chaos, and measuring its strength – the key to discovering just how far into the future predictions can be made. In a truly random phenomenon – such as choosing Lotto balls – there is no connection between each event: if, say,

a "17" has appeared in five consecutive draws, it is no more or less likely to turn up in the next one. With no link between past and future, there's no way of predicting what will happen next. At the other extreme are the regular systems like a quartz crystal watch, for which it's possible to predict what they will be doing far into the future.

But with chaotic phenomena like the weather, there is a link between past and future – it's just not very strong. As a result, knowing today's weather is helpful in predicting tomorrow's weather, but becomes useless in predicting what will happen next month. That's because the Butterfly Effect amplifies the smallest errors in data until they ruin any forecast.

The most common way of measuring the strength of chaos is to calculate the so-called Lyapunov timescale, which captures the rate at which errors grow over time. Computer models of the weather suggest a Lyapunov timescale of around 20 days: that is, even with the best data and fastest computers, it will never be possible to produce reliable forecasts more than around three weeks ahead.[1]

Studies of the apparently regular motion of the planets around the sun have revealed that they too behave chaotically, because of their gravitational fields pushing and pulling them in complex ways. Computer models suggest a Lypanuov timescale of around 100 million years for the Earth, after which its position becomes completely unpredictable.

While astronomers may lament the death of their dream of predicting the future, such vast time-scales have little practical importance. Far more significant is the discovery of chaos among asteroids, the chunks of rock which orbit mostly between Mars and Jupiter. Some of these asteroids are on orbits that bring them worryingly close to Earth, raising the prospect of a future impact – with potentially disastrous consequences. Identifying these asteroids and calculating exactly when they might strike is one of the most urgent tasks now facing astronomers. Yet the constant jostling caused by the gravity of the Earth and other planets has given some of these asteroids strongly chaotic orbits, with Lyapunov timescales as short as decades – ruling out any hope of forecasting their exact location more than a century or so ahead.[2] Clearly astronomers have no choice but to keep a close eye on these chaotic asteroids, to make sure they do not suddenly shift onto a collision course.

Chaos is not always bad news, however. If it can be found lurking in something that appears totally random, then it may be possible to predict future behaviour – even if only for a short time. Unsurprisingly, given the potential for making fortunes, much effort has been put into finding evidence for chaos in apparently random economic and financial data. And on the face of it, their dependence on a host of different factors should mean that the "random" behaviour of, say, gross national product or currency exchange rates really is chaotic – and thus potentially predictable.

Yet so far the evidence has been less than convincing. Studies by Dr Paul Ormerod of the London-based economic consultancy Volterra have shown that economic data appear to be more random than chaotic – which would at least explain the notorious unreliability of economic forecasts. Similarly gloomy conclusions have

---

Safety in numbers: ensemble weather forecasting

While chaos may have scuppered the dream of precise long-term weather forecasts, meteorologists have found ways of gauging the impact of chaos on a specific forecast – and thus how reliable it is likely to be. Known as ensemble forecasting, it involves making a set of forecasts, each one of which begins with slightly different initial conditions. If the atmosphere is in a state where the Butterfly Effect is strong, then the forecasts will diverge quickly, warning meteorologists not to put too much faith in them. The ensemble technique has shown that forecasts tend to be relatively unreliable in winter, especially around the British Isles.

---

been reached about financial data such as exchange rates. Even so, some City institutions are known to use mathematicians to spot bursts of predictability in financial markets – and if they are having success using chaos theory, they aren't saying, for fear of alerting their rivals.

While chaos may not be a sure route to wealth, it does seem important to health. Medical researchers have shown that a healthy heart does not beat with complete regularity, but instead is slightly chaotic. If the level of chaos becomes too big, however, the resulting "arrythmia" can be dangerous. Similarly, the electrical activity in the brain causing epileptic seizures comes in bursts lacking a healthy amount of chaos.

This has prompted the rise of a new theme in chaos research: finding ways of controlling it. The idea is to study the behaviour of, say, a beating heart and apply a stimulus, such as electrical signals, at just the right time to control the level of chaos. Professor William Ditto and colleagues at Georgia Institute of Technology in America have pioneered ways of controlling chaos in this way. Often the task is to turn chaotic behaviour into something more regular. In the case of epilepsy, however, the team has used what it calls "anti-control" methods to introduce a small amount of chaos into brain cell activity to dilute the concentrated bursts of activity associated with seizures.

The science of chaos has come a long way in the 40 years since Ed Lorenz's pioneering studies. So has the attitude of scientists towards it. Until the late 1980s, chaos tended to be dismissed as just another buzz-word, a trendy fashion seized on to explain everything from unreliable weather forecasts to cancer. Certainly not all of the grand claims made for chaos have stood the test of time. Even so, chaos is now recognised as a key feature of our world, affecting us all – even as we race to get to work on time.

Notes

1. The role of the Butterfly Effect in limiting predictive power may have been exaggerated, at least in the case of weather forecasts. Research by David Orrell at University College, London, has shown that the errors in weather forecasts simply don't follow the pattern predicted by the Butterfly Effect. If they did, the errors should grow exponentially with time, starting off relatively small but doubling in size every few days until they wreck the forecast accuracy. Yet the forecast errors of real computer models behave in a totally different way, rising rapidly in the first few days before levelling out (mathematically speaking, it increases with the square-root of time). Orrell has proved that this is precisely what one would expect if the main source of the error is not nature's Butterfly Effect at all, but simple inadequacies of the computer model. This opens up the prospect of dramatically better weather forecasts, because – unlike the Butterfly Effect – computer model error is something that can be fixed. To do it, meteorologists need to find better ways of capturing the subtleties of the Earth's atmosphere, from turbulence to cloud formation. Fixing these will reduce the basic errors made by the computer model in capturing the behaviour of the atmosphere, and thus cut the model errors which Orrell argues are chiefly to blame for forecasting errors in the first few days.

   It is not just weather forecasting that could benefit from this insight. Virtually every complex system, from colonies of cells to the global economy, contains the non-linear features which can lead to the Butterfly Effect. As such, Orrell's discovery could have implications for a host of phenomena whose behaviour seems to defy long-term prediction.

2. In 1995, Arthur Whipple of the McDonald Observatory at the University of Texas, Austin, published the results of his attempt to follow 175 Earth-crossing asteroids for thousands of years into the future. His results, published in the journal *Icarus*, showed that in many cases the whereabouts of these asteroids becomes unpredictable on timescales as short as a decade or so. Assuming that the initial position of such asteroids is known to a precision of around 100 km – which is pretty optimistic – the Butterfly Effect will boost that level of uncertainty to around the width of the Earth in less than a century. In other words, it is impossible to give assurances about whether such an asteroid will hit the Earth or not more than around fifty to a hundred years into the future.

Further reading

Chaos: Inventing a New Science by James Gleick (Minerva, 1996)

The Essence of Chaos by Edward Lorenz (UCL Press, 1993)

Newton's Clock: Chaos in the Solar System by Ivars Peterson (W. H. Freeman, 1993)

14
Cellular Automata

IN A NUTSHELL

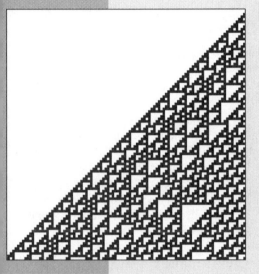

What do we need to create self-reproducing life? While this might sound like a question for biologists, the question has long attracted mathematicians, who have tried to pin down the absolute minimum number of ingredients needed to do the trick. In the 1940s, the brilliant Hungarian-American mathematician John Von Neumann put forward the idea of a self-replicating robot – an "automaton" – which had these characteristics.

It was, however, too theoretical to attract much interest, and progress only really began when Polish mathematicians Stan Ulam and Stan Mazur put forward a much simpler idea, based on grid-like patterns of cells whose behaviour was governed by simple rules. Now called Cellular Automata (CAs) these allowed scientists to simulate life-like behaviour using computers. CAs are now known to have two crucial abilities: they can act as universal computers and they can also self-replicate. This has led to some researchers claiming to have used CAs to make major breakthroughs in understanding not only living processes, but also the workings of the universe itself. The most vocal advocate of the power of CAs is the British mathematician Stephen Wolfram, whose recent book *A New Kind of Science* sets out a dazzling vision of CAs as a new tool for fathoming the wonders of the universe. So far, however, CAs have failed to make a big impact on science – and few scientists expect they will.

In May 2002, the world's media were proclaiming the dawn of a new era, a scientific revolution that would transform our understanding of life, the universe and everything. More amazing still, it was a revolution said to have been sparked by the work of a solitary genius working obsessively for over a decade. "Is this man bigger than Newton and Darwin?" mused one UK newspaper.

The man in question was Stephen Wolfram, a British physicist and software millionaire whose brilliance had given him legendary status while still in his teens. He published his first

TIMELINE

1929 Polish mathematicians Stanislaw Ulam and Stanislaw Mazur first begin speculating about the possibility of self-replicating robots.

1948 Hungarian mathematician John Von Neumann gives lecture outlining idea that life processes would appear in a machine exceeding a certain level of complexity

1952 Ulam tells Von Neumann of simple grid-like "Cellular Automaton" (CA) method for simulating life; Von Neumann unveils first-ever CA the following year.

1970 British mathematician John Horton Conway invents the Game of Life, a two-dimensional cellular automaton that produces "living" shapes.

1975 Italian PhD student Tommaso Toffoli revives interest in CAs by showing they can be run backwards in time like physical laws – hinting at their value for simulations.

1979 Physicist Christopher Langton claims to have created self-replicating artificial life on a computer using CAs.

1982 British student Stephen Wolfram begins to explore idea of CAs run on a computer as a new kind of science: neither wholly theoretical or experimental.

1985 Physicist Norman Packard uses CAs to simulate the formation and shapes of snowflakes, and later attempts to "breed" efficient CAs.

1986 Wolfram and a team led by Uriel Frisch show that CAs can be used to simulate the Navier–Stokes Equation, used to understand fluid flow (and the weather).

1992 Langton claims to show that certain types of CA are sufficiently complex to mimic life processes. His claims – and those of Packard – are challenged by others.

2002 Wolfram publishes *A New Kind of Science*, spelling out his claims for the importance of CAs. Many scientists remain very sceptical.

course as useless. He went to the California Institute of Technology, picked up a PhD and won a special fellowship for emerging geniuses.

By his early twenties Wolfram was tipped to make big advances in unravelling the mysteries of the cosmos. But when he unveiled his grand idea, it was not quite what many people expected. According to Wolfram, the key to the cosmos lay in grid-like patterns on a computer screen.

Such beliefs are usually a symptom of someone who has lost touch with reality. Yet coming from Wolfram, it was not a claim to be dismissed lightly, and some leading scientists thought he might be right – though they were demanding a lot more evidence. It finally came in 2002, in the form of a huge book with a grand title to match: *A New Kind of Science*.

The book was Wolfram's bid to be hailed as the discoverer of the power of those simple patterns to unlock the mysteries of nature. Yet he is far from being the first to have seen glimpses of something vast in those strange grid-like patterns that litter his huge book. Known as Cellular Automata (CAs), they were attracting the attention of some of the most brilliant minds of the twentieth century long before Wolfram was born.

Yet despite the great efforts made to unlock the potential of CAs, controversy still rages about their real significance. Are CAs truly the key to the cosmos … or just patterns whose beauty has led brilliant minds astray?

Their origins lie in coffee-house conversations in the Polish town of Lvov (now in the Ukraine) in 1929. Stanislaw Ulam and Stanislaw Mazur were two maths students at the local polytechnic, and regularly met to kick

research paper while still at school, and quit Oxford University a year after in the late 1970s, dismissing the physics around ideas. During one of these sessions, Mazur raised the question of what it would take to build a machine that could automatically make copies of itself. The pair pondered this "automaton" idea for a while, before moving on to other things. When they next discussed it, over 20 years later, their student musings would lead to the birth of CA research.

By that time one of the greatest mathematicians of all time had begun pondering the same question. Born in Hungary in 1903, John Von Neumann was already famed for his work in fields ranging from economics to computer science. In the 1940s he proved that computers needed a central processor, a memory, and an input/output device; this "Von Neumann architecture" dominates computer design to this day.

By the late 1940s, he wanted to do the same for life itself, identifying the key components and instructions needed to produce all the processes of life, including reproduction. The result would be the "automaton" that Mazur and Ulam had dreamt up years before.

As with the computer, Von Neumann identified the various components needed to create the automaton: a central controller, a copying device, a building device and a set of instructions. Today biologists recognise this as the checklist of components found in all forms of life; amazingly, Von Neumann had proved its importance years even before genes had been identified. Yet he was unhappy about the lack of practical detail about these components. How would one construct one of these automata?

Von Neumann mentioned this to his long-time friend and fellow-mathematician: Stanislaw Ulam. Recalling his coffee-house conversations over 20 years earlier, Ulam suggested beginning with something simpler than building a full-blown replicating automaton. Why not start with patterns of simple coloured squares on graph paper, each following a set of rules, and seeing if these patterns could make copies of themselves?

Von Neumann took up Ulam's suggestion, and set about creating a

JARGON BUSTER

Cellular automaton: a grid-like pattern of cells whose details are controlled by simple sets of rules. Each cell in the pattern takes on a variety of different states (ranging from simply "black" and "white" to far more complex varieties) according to the state of its immediate neighbours.

Von Neumann machine: A machine programmed with rules that allow it to self-replicate using components around it. Named after the eponymous Hungarian-American mathematician, Von Neumann machines show that all key processes of life are possible in a machine whose complexity exceeds a certain level.

Universal computer: A device that can perform any conceivable computation, from arithmetic to logical reasoning. Believed to be a key requirement of any system capable of mimicking life. Even the simplest Cellular Automata have been shown to possess the ability to perform universal computation.

Lattice gas: a way of simulating the complexities of fluid flow – for example, the percolation of gas through rock – using Cellular Automata (CAs) to mimic the behaviour of the gas molecules. The simple rules governing CAs allow computers to solve such problems, whose equations are notoriously complex.

The Brilliant Dr Von Neumann

Born into a wealthy banking family in Budapest in 1903, Janos – later anglicised to John – Von Neumann was a member of an astonishing group of minds who emerged in Hungary during the early part of the last century.[1] From the age of twelve Von Neumann was being taught by top mathematicians from Budapest University, and he published his first research paper at seventeen. By the age of twenty-two he was an assistant professor at Berlin University – the youngest person ever to hold such a post.

He then began laying the mathematical foundations of fields as diverse as Quantum Theory, Game Theory and the design of nuclear weapons. Bizarrely, however, Von Neumann is said to have regarded his work on cellular automata as his most important achievement.

"virtual creature" in the form of a grid-like pattern of "cells", each of which could take on any of twenty-nine states, according to a set of instructions. Made up of tens of thousands of such cells, the creature was designed to make copies of itself by changing the state of the cells around it to form the components Von Neumann had shown were needed for life. Eventually it would thus make a copy of itself – mimicking the key process of life: reproduction.

It was a bold idea, but the sheer complexity of the rules and states defeated Von Neumann: he never succeeded in building his virtual creature. After his death in 1957, his pioneering attempt to create artificial life was written up by a former collaborator, Arthur Burks at the University of Michigan. Searching for a name for what Von Neumann had tried to create, Burks hit on "Cellular Automaton" – now the standard term for these grid-like pattern of cells.

Yet while the field now had a name, Von Neumann had failed to build a CA, let alone explore its properties, and the field became dormant for many years.

It re-emerged in 1970, in unusual circumstances: an article in an American science magazine describing a weird game invented by a British mathematician. Called "Life", it had been devised at Cambridge University by John Conway, who had been exploring the idea that Von Neumann's virtual creature could be built using far fewer cells and states. Conway began with cells that came in just two forms – black and white – plus a set of rules that dictated the state of each cell according to the state of its eight neighbours.

After a huge effort, he found a set of rules that produced an astonishing variety of behaviour. While some CAs quickly "died", others flipped between two or more different states, while some patterns appeared to move across the grid like insects. Soon students with access to university computers were playing Conway's Game of Life, investigating the bizarre world of CAs for themselves.

Conway went further, however, showing that his CAs could act like a computer, performing logical operations needed to solve problems. Yet like Von Neumann, Conway failed to make either a self-reproducing automaton or a CA computer. He thought it could be done given a big enough grid, but hard proof was lacking. By the mid-1970s, CAs had once again slipped off the research agenda.

The flickering candle was kept alight at the Massachusetts Institute of Technology, where a team of researchers took a different – and grander – view of CAs. Ed Fredkin and his colleagues believed CAs could do more than look vaguely "alive". They saw them as the key to a whole new way of probing the secrets of nature. One of the team, Tommaso Toffoli, had shown that CAs had similarities to complex physical phenomena like turbulent fluids, where interactions between neighbouring regions is crucial.[2] Conway's Game of Life had shown that CAs could turn simple patterns into very complex ones at enormous speed. To the MIT team, this made CAs perfect for computer simulations of really complex physical problems like turbulence.

Their view was shared by Stephen Wolfram, who by then was at the prestigious Institute for Advanced Study in Princeton. During the early 1980s, he became the most vocal advocate of the view of CAs as a new way of doing science. According to Wolfram, many problems had stumped scientists because they had only two ways to attack them: experiment or theory. He believed CAs were a "middle way", allowing scientists to experiment with their patterns while searching for new theories for everything from turbulence to the patterning on sea-shells.

Wolfram began by focusing on CAs even simpler than those created by Conway: just black and white squares on a line. Yet he too found that some CAs started simple while others became astonishingly complex, and he classified them accordingly. Wolfram became convinced that the ability of even simple CAs to produce complex behaviour was clear evidence of their cosmic significance. By the mid-1980s, Wolfram's vision had succeeded in putting CAs back in the scientific limelight. His IAS colleague Norman Packard used CAs to simulate the growth of snowflakes, while a team led by the French physicist Uriel Frisch showed that CAs could be used to solve the Navier–Stokes Equation. Notorious for its complexity, this equation is

The ultimate cellular automaton: Rule 110

In *A New Kind of Science*, Wolfram examines the properties of hundreds of types of CAs, each of which evolves according to a different set of rules. Out of the 256 specific varieties he considers, he focuses particular attention on one, which obeys "Rule 110". This particular CA, says Wolfram, is special because it has been mathematically proved to exhibit "universality" – that is, an ability to simulate any feature of any system. As such, it exemplifies the feature of CAs which, according to Wolfram at least, makes them so exciting: their paradoxical combination of simplicity and profoundly complex behaviour. Quite how useful this will prove in real-life simulations remains, however, unclear.

central to understanding fluid flow – and is vital in everything from weather forecasting to aircraft design. At the University of Michigan, Chris Langton even claimed to have identified the conditions needed to create "living" CAs, and to have found the holy grail: a self-reproducing CA.

Despite this flurry of activity, most scientists remained unimpressed. To them, CAs remained at best intriguing, but more likely a complex but meaningless pastime. By the late 1980s, Wolfram had moved on to other things, and the field yet again lost its lustre.

Wolfram himself never quite gave up on CAs, however, and after setting up a multi-million-pound software business, he began a decade of intensive, solitary research into their properties that resulted in *A New Kind of Science*. It became an instant best-seller, yet among scientists the response was far from enthusiastic. Many accused Wolfram of failing to credit the work of others and indulging in hype. They also pointed out that there is nothing new or unique about how CAs produce complex behaviour from simple rules: other ways have been known for years. And few thought Wolfram's book had proved his claim that CAs are the basis for "a new kind of science".

Are CAs as important as Wolfram – and many other brilliant minds before him – believe? The fact is that the history of CAs is littered with grand claims about CAs, and yet so far they have failed to have any significant impact. Wolfram's massive book may be the spark that finally triggers a CA revolution. Yet on past form, it is far more likely to prove the headstone on the grave of a Big Idea that never delivered.

Notes

1. The roll-call of geniuses that emerged from Hungary at the same time as Von Neumann included Leo Szilard, discoverer of the "chain reaction" of nuclear fission, the aerodynamicist Theodore Von Karman, the father of the H-Bomb Edward Teller, the mathematician Eugene Wigner and three science Nobel Prize winners, including Denis Gabor, who invented the laser hologram. Historians have suggested that part of the reason for the emergence of so many influential minds in so short a time lies with the reorganisation of Hungarian school system around the end of the nineteenth century. It was led by Mór Kármán, a professor of philosophy and education (and father of Theodore), who modelled schools on the German gymnasium system, and put special emphasis on encouraging the most brilliant students.

2. Toffoli is credited with having wrested CAs from pure mathematicians and highlighting their potential use in computer simulations of real-life phenomena. In 1977, he produced a mathematical proof that CAs possessed "reversibility" – that is, an ability to be wound backwards in time to get back to their starting-point. This suggested that CAs could prove useful in simulating natural phenomena, and gave the field a life beyond the theoretical investigations of pure mathematics. Toffoli himself went on to build hardware specially tailored for ultra-fast CA-based simulations.

Further reading

A New Kind of Science by Stephen Wolfram (Wolfram Publishing, 2002)

Available online at www.wolframscience.com/nksonline/toc.html

Artificial Life: The Quest for a New Creation by Steven Levy (Penguin, 1993)

Various online simulations of CAs are available at www.collidoscope.com/modernca/

15
Extreme Value Theory

IN A NUTSHELL

From hurricanes to high-jump records, every so often a freak event quite unlike anything seen before can appear seemingly out of the blue. But just how outlandish are such events? That is the question answered by Extreme Value Theory, which takes existing records of, say, the worst floods over the last 500 years, and uses them to predict the chances of seeing an even worse flood in the future.

Developed by mathematicians in the 1920s, EVT was long regarded with suspicion because of its seemingly magical ability to predict even unprecedented events. Now there is growing confidence in its abilities, which are being put to use in areas as diverse as financial risk planning and marine safety. One major use is in the insurance industry, where companies apply the formulas of EVT to estimate the likelihood of major disasters, and ensure there are enough funds to cover the cost.

These are strange times. All over the world, records that have stood for centuries are being utterly blown away. In just one week in May 2002, an unprecedented storm of 300 tornadoes ripped across America's mid-west, causing over £1 billion worth of damage. In August 2003, Europe sweltered in a heat wave which saw Gravesend in Kent record the highest ever UK temperature of 38.1°C, while over 14,000 died of the effects of heat in France. Even the sun joined in the record-breaking spree, releasing the largest ever flare in November, and forcing aircraft to alter course to minimise cosmic radiation.[1]

What on earth can we expect next? Can we do anything to protect ourselves from what Nature may now throw at us? Extreme events often prompt such questions. They reflect our dread of what the future might hold, and our feelings of impotence to act to prevent a repeat of past disasters.

With the world springing ever bigger surprises on us, there is now growing excitement over a technique with the power to do the seemingly impossible. Known as Extreme Value Theory (EVT), it can give insights into extreme events that lie years, decades or even centuries in the future.

TIMELINE

1709 Swiss mathematician Nicolas Bernoulli first raises concept of extreme values in a simple geometrical problem concerning dots on a line.

1928 Cambridge mathematician R.A. Fisher, together with L.H.C. Tippett, develop the basic mathematics of Extreme Value Theory.

1953 Severe storm surge breaches the sea defences of the Netherlands, killing 1800. EVT is used to check the design of the new defences.

1958 Emil Gumbel at Columbia University, New York, publishes first textbook on EVT, bringing its power to the notice of many outside academia.

1993 Chinese athlete Wang Junxia smashes world record for 3000 metres by 17 seconds. EVT helps defuse allegations of drug abuse.

1994 Laurens de Haan and Karin Aarsen at Erasmus University, Rotterdam, use EVT to estimate the maximum human lifespan to be 124 years.

1995 Collapse of Barings bank and discovery of fraud at Japan's Daiwa bank triggers concern about standard methods for minimising risk.

1995 Financial debacles prompt Alan Greenspan, chairman of the US Federal Reserve, to talk of the potential benefits of using EVT in risk assessment.

1996 Study of Britain's sea defences using EVT leads to warnings that they are much too low to cope with storms in the foreseeable future.

1998 Long Term Capital Management, one of America's biggest hedge-funds, goes to brink of collapse, and highlights need for better risk assessment.

2000 Mathematicians at Lancaster University use EVT to help investigators solve the riddle of the sinking of the *Derbyshire*, and to make safety recommendations.

Around the world, governments are now turning to EVT to gauge the likelihood of extreme events, and to prepare defences against them. In the financial world, EVT is being used to assess the risk of natural disasters, and to ensure institutions have the financial reserves needed to cover the likely impact. The technique is being used to protect ships from the most extreme storms they may face while plying the world's oceans. It has even helped biologists cast light on the maximum lifespan of humans.

In each case, EVT takes records of past extremes and extracts insights into what might happen in the future – even putting figures on the chances of events that have never occurred before. It sounds almost miraculous; perhaps not surprisingly EVT is still regarded with suspicion in some quarters. Yet advocates of the method point out that it is based on sound mathematical foundations, unlike most of the rules of thumb so often used to peer into the future.

Glimmerings of what has since become EVT have been traced to the early eighteenth century, and some musings by the Swiss mathematician Nicolas Bernoulli. Yet it was not until the 1920s that the idea of predicting the unprecedented first attracted serious attention. At its heart lies the concept of the "distribution", a mathematical formula giving the relative frequency of a particular quantity. For example, the chances of observing people with a particular height follows the famous bell-shaped curve known as the Gaussian distribution. This shows that most people have near-average height – which gives the "hump" in the centre of the distribution. The long, thin "tails" of the distribution reflect the fact that the chances of meeting extremely short or tall people are much lower.

The Gaussian distribution applies to a host of phenomena, and early attempts to predict the probability of extreme events were based on calculations involving its "tails". But during the 1920s, mathematicians made clear their qualms about the resulting predictions, and whether the Gaussian curve could be trusted to deal with extreme events. In 1928, the brilliant Cambridge mathematician R.A. Fisher and his colleague L.H.C. Tippett launched what became known as Extreme Value Theory with a paper showing that extreme events do indeed follow their own special types of distribution. While their precise mathematical form differs, each of the distributions reflects the intuitive idea that the more extreme an event is, the less likely it is to occur. Just how much less likely could be gauged by fitting the distributions to data on past extremes – for example, the highest coastal storm surge seen each year – and then projecting forward into the future. The formulas of EVT then spat out estimates for the risk of storm surges of a given size being seen, including ones so extreme they had never been seen before.

Despite its obvious practical value in so many fields, EVT was regarded with suspicion for many years. During the 1940s, two events took place which helped overcome those doubts. First, the formulas were put on a rigorous mathematical footing by the Soviet mathematician Boris Gnedenko. Then the German-born American mathematician Emil Gumbel began applying EVT to the problem of predicting flood levels from past records, with great success.

The importance of this application of EVT was brought into stark focus in February 1953, when a huge storm surge off the Dutch coast broke through centuries-old sea defences. The inundation killed 1800 and destroyed 47,000 homes, and led to demands for action to prevent it happening ever again. A panel of experts set about studying past records of floods to gauge the size of the problem, and found that the 1953 event was far from being the worst ever. On All Saints Day in November 1570, the country had been one of several in Western Europe devastated by a storm surge in which the height of the sea increased by over 4 metres – at least

JARGON BUSTER

Gaussian distribution: The familiar bell-shaped curve showing the relative frequency of, say, the heights of a large group of people. Its shape reflects the fact that most people have heights close to the average, while very tall – or very short – people are relatively rare.

Extreme value distribution: A curve which does for extreme values what the Normal distribution does for run-of-the-mill values. Based on past records of, say, the highest values of monthly rainfall, the extreme value distribution produces a curve that can give estimates of the chances of having even higher levels of rainfall in any given year.

Time series: A collection of data stretching back over time, from which an extreme value distribution can be extracted. For example, by studying historical records of the highest river levels reached over several centuries, planners can estimate the height of flood defences needed to protect homes from the worst excesses

expected over the coming centuries.

Value-at-risk: The biggest loss likely to hit a financial institution over a given amount of time, with a certain probability. A big concern for insurance companies, estimating the "VaR" is now a major application for EVT.

Gaussian distribution

Named after the eighteenth century German mathematician Carl Gauss (though discovered earlier by the French mathematician Abraham de Moivre), this bell-shaped curve shows the prevalence of properties – such as IQ, height or weight – which have an average value, but are subject to random variations. The central "hump" shows that the average value is the most common, while the sloping sides reflects the existence of smaller numbers with more unusual values both higher and lower than the average.

15 cm above the 1953 event. An estimated 400,000 people lost their lives.

The challenge facing the expert panel lay in deciding just how high to build the sea defences. Set the level too high, and building the sea-walls would be unnecessarily expensive; set it too low, and the country could face a repeat disaster all too soon. The panel concluded that coastal defences around 5 metres above sea level should protect the country for thousands of years to come. But how reliable was their estimate? To find out, a team led by Laurens de Haan at Erasmus University, Rotterdam, used EVT to estimate the risk of a flood exceeding the 5-metre level proposed by the panel. Analysis of the historical flood data allowed the team to find an EVT distribution that accounted for past extreme floods – and to extrapolate into the future. Their calculations showed that the panel's recommendations could be expected to protect the country for many centuries to come.

EVT is now widely used to assess the risk of flooding and the adequacy of defences against it. Even so, it has to be used with care. In 1996, mathematicians at Lancaster University found subtle flaws in the way EVT had been used to design Britain's 800 km of sea defences. The team concluded that some parts of the defences would need

to be two metres higher to protect coastal areas from serious flooding.

Not surprisingly, insurance companies have been among the first to exploit the power of EVT. For many years, actuaries gauged the likely risk posed by various forms of disaster by using empirical rules of thumb such as the "20–80" rule, which states that 20 per cent of the severe events account for over 80 per cent of the total payout.[2] In the mid-1990s financial mathematician Paul Embrechts and colleagues at the Swiss Federal Institute of Technology (ETH) in Zurich, decided to check the validity of such rules with EVT. They found that the "20–80" rule does work well for many insurance sectors – but it can also fail very badly. For example, using EVT to study past data on claims, the team found that a "0.1–95" rule applies to hurricane damage. In other words, the real threat comes from the 1-in-1000 storm, which can devour 95 per cent of the total payout in one go. Such discoveries are helping insurance companies optimise their risk coverage, broadening the range of threats they can cover at sensible premiums – with benefits to both the companies and their clients alike.

Around the same time, a series of financial disasters sparked huge interest in EVT among quantitative analysts, the so-called "rocket scientists" who use sophisticated mathematical methods to create profits and avoid losses – or, at least, attempt to. In February 1995, the Singapore subsidiary of Barings, the world's most famous merchant bank, lost almost £900 million through the dealing of a single trader named Nick Leeson.[4] Just a few months later, the Daiwa Bank of Japan discovered a £700 million hole in its accounts through the activities of another rogue trader, Toshihide Iguchi. It was an unprecedented double-whammy, and prompted Alan Greenspan, chairman of the US Federal Reserve, to talk of the potential benefits of using EVT in financial risk assessment.

This was the kind of advice that could have helped avoid an even bigger financial catastrophe that struck three years later. In the summer of 1998, Russia's economy went into meltdown, and its government did the unthinkable: default on the nation's domestic debt repayments. It was an unprecedented move that took everyone by surprise – especially Long Term Capital Management, a huge American hedge-fund company, which found itself caught out to the tune of £100 billion. LTCM was saved by a bail-out organised by major banks, but it had been a shocking experience: the company worst-case scenario was only 60 per cent as bad as what actually took place.

The debacle focused concern on how financial fund-managers estimate their risks. It later emerged that LTCM had been using methods which assume losses follow the bell-shaped Gaussian distribution. But as mathematicians had warned over 70 years earlier, the most extreme losses follow a different distribution – and one likely to vastly underestimate the true losses. Had LTCM followed Greenspan's advice and used EVT methods to assess its risk, it may well have done a better job of protecting itself from the unthinkable.

Awareness of the power of EVT is now spreading into ever more fields, from human biology to marine engineering. At Erasmus University,

When are results just too good to be true?

Every four years athletes from across the world attempt to break records at the Olympic Games. But sometimes records aren't just broken – they're utterly smashed by performances quite unlike anything seen before. But can a performance be too good to be true? That is a question that can be addressed using Extreme Value Theory.

Take the case of Wang Junxia, China's middle-distance running star of the 1990s. At the national championships in Beijing on 13 September 1993, she ran the 3000 metres in an astonishing time of 8 minutes 6 seconds – smashing the previous record set nine years earlier by an astonishing 17 seconds. Was it just an amazing performance, or had she cheated? Drugs tests proved negative, but to this day many commentators claim Wang must have cheated. Mathematicians Michael Robinson and Jonathan Tawn at Lancaster University decided to see what light EVT could cast on the controversy. Using past records, they calculated the most extreme – that is, quickest – time plausible according to current trends. The analysis showed that, while Wang's record was amazing, it was still 3 seconds slower than the fastest time plausible on the basis of previous records.

Rotterdam, Laurens de Haan and his colleagues have used EVT to tackle the mystery of the maximum lifespan of humans. By analysing life span records for the "oldest old", they have found evidence that the ultimate, most extreme lifespan of humans is around 124 years. And so far, it seems to be true: the oldest person with authenticated records was Jeanne Calment of France, who died in 1997 aged 122.[3]

The methods of EVT are also being used to solve more tragic mysteries. In 1980, the giant bulk-carrier *Derbyshire* sank with all forty-four on board in a typhoon off Japan. For years there were questions about whether the ship was a victim of flawed design, or whether poor seamanship was to blame, as a public inquiry concluded in 1997. Three years later, a second inquiry exonerated the crew, after EVT experts Professor Jonathan Tawn and Dr Janet Heffernan at Lancaster University helped uncover the real cause: a freak wave so violent that it smashed through the ship's forward hatch. The research has led to recommendations that hatch strengths be substantially increased to cope with such events.

As a result, mariners have joined the millions of other people around the world whose lives have been made safer by the esoteric mathematics of Extreme Value Theory.

Notes

1. Extreme events in deep space can affect life on Earth. In November 2003, the sun blasted our planet with the biggest flare ever recorded, with particles hitting the upper atmosphere at over 8 million km/hr. Aircraft routes had to be changed to dodge the increased cosmic radiation.

2. Insurance companies have long relied on simple rules of thumb to gauge the likelihood of taking huge hits from disasters. Extreme Value Theory is now helping to find more reliable rules for specific threats, such as tropical storms.

3. Jeanne Calment of France made headlines in 1995 when she became the oldest person in history, at the age of 120. Two years later she died, aged 122. According to Extreme Value Theory, humans may be able to live even longer.

4. Rogue trader Nick Leeson was responsible for one of the most notorious financial extreme events in history. A series of bad trades at Barings, the merchant bank, led him to rack up losses of almost £900 million.

Further reading

Extreme Values in Finance, Telecommunications and the Environment, edited by Bärbel Finkenstädt and Holger Rootzén (Chapman & Hall, 2003)

Extreme Value Distributions, by Samuel Kotz and Saralees Nadarajah (World Scientific, 2000)

www.nottingham.ac.uk/~lizkd/evvar.doc

http://mathworld.wolfram.com/ ExtremeValueDistribution.html

http://www.library.adelaide.edu.au/ digitised/fisher/63.pdf

IN SICKNESS AND IN HEALTH

16
Evidence-based Medicine

IN A NUTSHELL

For centuries doctors made life-or-death decisions about how best to treat their patients on the basis of little if any scientific evidence. Tests of new treatments were typically carried out on small numbers of patients, with results that were often contradictory. The emergence of statistical techniques for making sense of experiments led in the 1940s to randomised controlled trials (RCTs), in which patients are randomly chosen to receive the new treatment, or the existing one (or just a placebo). The first such trials, carried out by UK Medical Research Council scientists, were so successful that RCTs became the core of a radical new movement known as Evidence-based Medicine (EBM).

EBM has helped to give doctors clear, objective evidence about which treatments work, and which do not – with huge benefits for patients. Even so, critics point out that it has failed to end arguments over which treatment is best for individual patients.

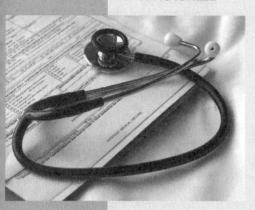

As the most distinguished consultant at St Swithins Hospital, Sir Lancelot Spratt did not like having his judgements challenged. If he declared that a patient should be treated with a specific drug, that was it – end of argument. Woe betide any junior doctor who suggested that some new drug might be better.

Sir Lancelot may be only a fictional character in the 1950s comedy classic film *Doctor in the House*, but his manner captured the way life-or-death medical decisions were made at the time (not least because his creator, Richard Gordon, was a former hospital doctor). Imperious consultants would draw on years of personal experience, knowledge and insight to reach what they at least regarded as decisions of divine infallibility.

Yet while such consultants undoubtedly had vast levels of expertise, few seemed to realise just how much of their knowledge was past its sell-by date. The post-war years marked the start of an explosion in medical research, with a host of new therapies for diseases ranging from tuberculosis to leukaemia emerging from laboratories.[1]

What medical researchers urgently needed were ways of comparing

TIMELINE

1747	Scottish naval physician James Lind carries out trials of various treatments for scurvy on a dozen sailors, and identifies citrus fruit as a remedy.
1925	Statistician Ronald Fisher publishes *Statistical Methods for Research Workers*, setting out techniques for testing the "significance" of new research findings.
1943	UK Medical Research Council (MRC) begins investigation of patulin treatment for the common cold in a pioneering "double blind" controlled trial.
1946	Austin Bradford Hill persuades the MRC to set up randomised controlled trials (RCTs) of whooping cough vaccine and streptomycin therapy.
1967	Canadian biostatistician David Sackett sets up department at McMaster University which becomes the home of the Evidence-based Medicine movement.
1976	Statistician Gene Glass at Arizona State University coins the term "meta-analysis" for techniques used to draw conclusions from collections of trials.
1993	The UK Cochrane Collaboration – named after epidemiologist Archie Cochrane – set up as a worldwide centre for assessing the latest evidence on medical therapies.
1995	Major trial reveals magnesium injections to be useless for heart attack victims – contradicting results from a meta-analysis of many small trials.
1998	Controversial Cochrane review of evidence from clinical trials of widely used albumin treatment for burns victims shows that it is actually worse than useless.[2]
1999	UK National Institute for Clinical Excellence (NICE) set up to ensure the National Health Service uses therapies backed by best available evidence.
2003– present	Evidence mounts that pharmaceutical companies suppress inconvenient data,[3] raising risk of misleading conclusions.

these therapies and assessing their effectiveness. And what patients needed were consultants and GPs who knew about these breakthroughs – and based their choice of therapy on more than old prejudices and bluster.

These demands have led to the emergence of one of the most influential – and controversial – medical developments in modern history: Evidence-based Medicine (EBM). Yet behind that innocuous-sounding term lie concerns which have provoked intense argument among doctors and medical researchers alike.

Not surprisingly, the debate focuses on the reliability of the evidence needed to practise EBM. According to its advocates, this involves making clear and conscientious use of the best current evidence available about, say, cancer treatment. This typically means using evidence drawn from clinical trials, which compare patients treated with, say, a new leukaemia drug with those given a standard treatment, or even a useless placebo.

In theory, such trials can provide clear evidence for the effectiveness, or otherwise, of life-saving treatments. But there is mounting concern that such trials aren't all they seem. Many doctors suspect the results are often misleading, because the patients in trials are hand-picked and thus unrepresentative of typical patients. Trials are also often funded by huge pharmaceutical companies – raising fears that the results are massaged to give the "right" answer, or even suppressed.

Even so, many doctors insist that EBM is far better than the old ways of making medical judgements – and has led to some spectacular successes. As long ago as 1747, the Scottish physician James Lind[4] used a simple clinical trial to tackle the scourge of the world's navies: scurvy.

Since the days of Columbus, doctors had noticed that sailors on long voyages fell victim to a disease which began with swollen gums and loose teeth, then lassitude, haemorrhage – and death. By Lind's day scurvy was claiming the lives of up to 40 per cent of Royal Navy sailors on long-range voyages – far more than died from battle wounds.

Lind was convinced the cause lay with the poor diet of the sailors, and in May 1747 he began a study that would change the course of British history. He found six pairs of patients with closely matched symptoms, and gave them identical diets – along with treatments he thought might alleviate their symptoms, ranging from quarts of cider to strong mouthwash. Within a week one treatment had proved astonishingly effective: a daily ration of oranges and lemons. Both patients given the fruit completely recovered from their lethal affliction.

Lind's pioneering use of EBM had revealed the first effective treatment for scurvy (which we now know to be due to lack of vitamin C). Yet the confused account by Lind and paucity of data failed to convince the Admiralty, which took another 40 years to make citrus fruit part of the naval rations. Once introduced it had a dramatic effect on both the health of crews and on history: at the Battle of Trafalgar in 1805, the Royal Navy was all fighting fit, while the French had to leave one-third of their ships in port because their crews were sick with scurvy.

Despite this clear demonstration of the value of EBM, the idea of setting up trials and collecting data as objectively as possible failed to catch on among medical researchers – who mostly lacked the mathematical skills needed to analyse the results. The kind of techniques required began to emerge in other fields first, such as agricultural science. In 1925, the British mathematician Ronald Fisher, one of the founders of modern statistics, gave scientists a set of tools for making sense of data, in a landmark book entitled *Statistical Methods for Research*

JARGON BUSTER

Meta-analysis: A mathematical technique pioneered by the Victorian Astronomer Royal Sir George Airy, in which the results of data from separate, but similar, trials are combined to mimic the effect of a single, huge trial.

Bias and confounding: Two key effects that can lead medical researchers to draw the wrong conclusions from medical trials. Bias is a systematic skewing of results away from the truth caused by, for example, journals refusing to publish "boring" negative results. Confounding is the effect on a trial result of some factor – e.g. smoking or social status – that the researchers failed to take into account when designing trials.

Double-blinding: A procedure designed to protect medical trials from hidden bias by ensuring neither the patients nor the doctors treating them know who is receiving the new therapy, and who is being given the placebo or the old-style treatment.

Randomised controlled trial: Regarded as the gold standard of Evidence-based Medicine, an RCT is a study of some new therapy where patients are randomly allotted to two groups called "arms": those chosen to receive the therapy in the "treatment arm", and those given the existing therapy or a placebo in the "control arm".

Oranges and lemons ... and scurvy

For centuries the symptoms of scurvy struck dread into the hearts of sailors. The first signs seemed trivial: aches and pains, and a feeling of tiredness. But then the unmistakable signs began to appear: tiny blood-blisters and blotches on the legs, swelling gums, and loose teeth. Without treatment, the lassitude turned into unconsciousness – and death.

Citrus fruit both prevented and cured the disease by supplying ascorbic acid – "Vitamin C" – which is needed for the biochemical reaction that creates collagen, the protein that forms the connective tissue and bone that hold the body together. Without Vitamin C, collagen is not robust enough for its crucial role, and it starts to disintegrate, with predictably lethal results.

Workers. They included "significance tests", designed to show whether, say, differences between patients given a new drug and those left untreated were truly convincing.

A major breakthrough came in the 1940s, when the UK Medical Research Council set up trials incorporating concepts now at the heart of Evidence-based Medicine: double-blinding and randomisation. In 1946, the MRC began "double-blinded" trials of a whooping cough vaccine and the use of the antibiotic streptomycin for treating tuberculosis – that is, neither the patients nor the doctors treating them knew who was getting which treatment, in an attempt to prevent patients or their doctors fooling themselves into seeing effects that weren't real. Following recommendations by the British medical statistician Austin Bradford Hill, the choice of which patients were treated and which were not was also made randomly. This helped cut the risk of patients getting the treatment being unusual in some way, thus giving a false impression of the drug's effectiveness.

The success of these studies in demonstrating the efficacy of the vaccine and antibiotic led to such "randomised controlled trials (RCTs)" becoming the gold standard test for new medical treatments. RCTs are now at the heart of evidence-based medicine, providing doctors with hard, scientific evidence on which to base their decisions.

Or at least, so it seems; the reality is proving to be more complex. No sooner had RCTs entered the mainstream than fears emerged that many trials included too few patients to show whether a drug was really useful. The result was a slew of "false negatives", in which small trials failed to reveal the *true benefit of drugs.*

During the 1980s, medical researchers began using a technique developed years earlier by physicists that promised to combat the problem of small trials. Known as meta-analysis, it allowed the results from many such trials to be combined, boosting their ability to detect even small benefits – which could nevertheless save thousands of lives. In 1985, researchers applied meta-analysis to over thirty small trials of a drug called streptokinase, used in the treatment of heart attacks. Most of the trials had failed to detect any benefit, yet the meta-analysis revealed the drug could save the lives of almost a quarter of patients treated with it.[5]

By the end of the 1980s, the wealth of insights emerging from RCTs and meta-analysis threatened to overwhelm the very people trying to make sense of it all. Among them was a former obstetrician named Iain Chalmers, who decided to set up a centre in Oxford dedicated to reviewing and summarising the latest data on medical therapies in a way doctors could act on. Launched in Oxford in 1993 as the Cochrane

Collaboration, it quickly became a central part of the EBM movement, which by then was sweeping the medical profession.[6]

But just two years later, advocates of EBM were reeling from the findings of a study that still casts a shadow over the whole concept. During the early 1990s, a meta-analysis of small trials had led doctors to hail magnesium sulphate as an "effective, safe, simple and inexpensive treatment" for heart attack victims. But in 1995, a huge RCT involving 58,000 patients concluded that magnesium sulphate was useless. What had gone wrong – and what were doctors to believe?

Sifting through the evidence, statisticians made some worrying discoveries. There were signs that the meta-analysis had failed to include studies showing negative results – the suspicion being that researchers and journals simply did not bother publishing "bad news". On the other hand, the huge RCT appeared to include relatively healthy patients who had been treated relatively late – which would make the treatment seem less effective than it really was.

In 2002, the results of another major RCT again failed to find any benefit from magnesium. Yet while this should have ended the argument, some experts still insist that the results are still not conclusive.

The debacle over magnesium sulphate highlighted an awkward fact about Evidence-based Medicine: despite the use of apparently "objective" methods like RCTs, medical experts can still wrangle

---

The curse of "underpowered" studies

Medical scientists regard double-blind randomised controlled trials (RCTs) as the ultimate test of whether a new therapy works or not. Yet even RCTs can produce very misleading results if they fail to include enough patients.

Studies have revealed that many trials reporting negative results are actually "underpowered", meaning they are simply too small to detect potentially important medical benefits.

One of the most common victims of underpowered studies are alternative and complementary therapies like acupuncture. With no huge multinational interested in funding big trials, such therapies are often studied in small, low-cost trials backed by charities – which, almost inevitably, fail to find compelling evidence.

---

endlessly over the results and their reliability. In 1998, researchers working for the Cochrane Collaboration made headlines with a study suggesting that a standard treatment for burns victims was actually worse than useless, killing more patients than it saved. Other studies failed to confirm the risk, and many hospitals still use the treatment.

Evidence is also emerging that some researchers hand-pick patients to take part in RCTs, while others trawl through raw data looking for positive results more likely to attract interest from medical journals. Many trials also fail to make clear their source of funding – despite evidence showing that trials backed by drugs companies are much more likely to claim positive results than independent studies.

Since Lind's pioneering work on scurvy, the methods of Evidence-based Medicine have saved the lives of countless people. While they are no panacea, few would want a return to the bluster of Sir Lancelot Spratt.

Notes

1. The last 50 years have been a golden age of drug discovery. Antibiotics like streptomycin defeated diseases like TB, while levodopa and chlorpromazine helped with afflictions like Parkinson's disease and schizophrenia. Meanwhile, anti-cancer drugs like vincristine have boosted childhood leukaemia survival rates a thousandfold.

2. As the most abundant protein in blood, albumin plays a key role in maintaining fluid pressure in cells, and allowing the flow of vital compounds in and out of cells. As burns victims often suffer massive fluid and protein loss, doctors believed it made sense to replace the albumin. However, this practice has been thrown into doubt by recent studies.

3. The reluctance of researchers and journals to publish negative findings is known as publication bias – and it can have a major impact on attempts to gauge the true effectiveness of a drug. Statisticians have devised ways of detecting publication bias, but compensating for the falsely optimistic view of a drug's effectiveness created by this bias is far from simple.

4. Born in Edinburgh in 1716, James Lind was just a lowly surgeon's mate when he carried out his famous comparative study of remedies for scurvy in 1747. The following year Lind began formal medical study, publishing his research into scurvy in 1753. Lind's failure to convince the Admiralty of the benefits of citrus fruit was not wholly due to bureaucratic stupidity: his research

results were confusing – and in any case, not even Lind could explain why citrus fruit worked. It took over 50 years for the case to be made, with citrus fruit being included in the Royal Navy rations in 1795 – the year after Lind's death.

5. One of the wonder-drugs of the post-war era, the enzyme streptokinase has saved the lives of thousands of people by dissolving potentially fatal clots circulating in their bloodstream. Streptokinase works by combining with plasminogen, a constituent of blood plasma, to form an enzyme which attacks fibrin, the fibrous material which forms the basis of blood clots.

6. The power of Evidence-based Medicine is reflected in the logo of the Cochrane Collaboration which shows the results of trials of corticosteriods given to women about to go into premature labour. Positive results emerged as early as 1972, yet the lack of a Cochrane-style organisation meant most doctors did not know – leading to tens of thousands of preventable deaths.

Further reading

www.people.virginia.edu/~rjh9u/scurvy.html

www.cochrane.org

www.nice.org.uk

www.cebm.net/background.asp

17
Epidemiology

IN A NUTSHELL

Over the centuries, physicians have suspected links between medical conditions and factors such as sanitation, occupation or diet. Proving the reality of such links is often very difficult, however – as there are usually many alternative explanations. Epidemiology began around 150 years ago with simple descriptions of the whereabouts and conditions of those afflicted by certain diseases, and maps revealing locations of high prevalence. In the early twentieth century, these descriptive methods began to give way to sophisticated ways of comparing groups of people to reveal the risk factors involved in specific ailments, or the common factors among those free of disease. Statistical methods helped assess the chance of findings being mere flukes, or the result of some other hidden cause. Today epidemiology is relied on to probe a host of issues, from the dangers of mobile phones to the benefits of drinking red wine.

Every year millions of tourists visit the north-west region of Wales to enjoy the spectacular countryside of Gwynedd. With its world-famous Snowdonia National Park and beautiful coastline, it looks idyllic – a part of the world untouched by the troubles of our time.

The reality is very different. Astonishingly, the region is still affected by the aftermath of the Chernobyl nuclear disaster of 1986, whose fall-out blighted this particularly rugged and rainy part of Europe. So far, the effects have been felt chiefly by Gwynedd's hill-farmers, whose livestock are under movement and sale restrictions. But early in 2004, a far more worrying possibility reared its head. Official statistics revealed that those living in the region have above-average rates of cancer.

Coming 18 years after the Chernobyl incident, there is understandable fear that these cancers are a direct result of the fall-out. But while understandable, is such fear really justified? To find out, scientists are using the techniques of one of the most important – and controversial – of modern scientific disciplines: epidemiology.

Literally the study of epidemics, the Greek roots of the word – "among the people" – capture the broader truth

TIMELINE

1775 The English surgeon Percivall Pott notes the high prevalence of scrotal sores among chimney-sweeps, and concludes it is cancer caused by soot.

1842 Publication of *Sanitary Conditions of the Labouring Population* by Edwin Chadwick reveals link between living conditions and disease.

1846 Study of a measles epidemic in the Faeroe Islands by the Danish doctor Peter Panum reveals protective effect of previous infection, and its contagious nature.

1854 John Snow uses simple epidemiological methods to identify a water-pump in London's Broad Street as the source of the capital's killer cholera outbreak.

1912 Janet Elizabeth Lane-Claypon publishes pioneering "cohort" study, with a statistical comparison of the health effects on babies of breast milk versus cow's milk.

1948 Scientists from US National Institutes of Health set up study of citizens of Framingham, Massachusetts, to uncover causes of heart disease.

1950 Richard Doll and Austin Bradford Hill begin comparing lung cancer rates between smokers and non-smokers, showing cigarettes to be the cause.

1981 Epidemiologists in New York and Los Angeles describe puzzling rise in rare forms of pneumonia and cancer among gay men – presaging the global Aids epidemic.

1989 Studies of the records of 17,000 children suggests allergies may be triggered by inadequate exposure to viruses, launching the "Hygiene Hypothesis".

1997 Researchers at St Bartholomew's Hospital, London, use meta-analysis to reveal cancer and heart disease risks caused by so-called passive smoking.

2003 Worldwide alert follows the discovery of a new form of virus that emerged in China, known as the Severe Acute Respiratory Syndrome (SARS) virus.

that epidemiology is the study of health effects of groups of people. Though its origins lie in the understanding of infectious diseases, it has come to encompass far more, from identifying the cancer risk from cigarettes to probing the health effects of vegetarian diets.

This is often a tough challenge, involving careful follow-up of the health of thousands of people over many years. There are also many pitfalls. An apparent rise in cancer rates can be due to simple chance – a possibility that can only be ruled out using sophisticated statistical analysis. In Gwynedd, for example, the rate of rectal cancer is 50 per cent higher than in the rest of Wales – which sounds dramatic, but actually amounts to just one extra case per 10,000 people.

As well as investigating whether these extra cases could be due to fluke alone, scientists assessed their inherent plausibility. The rectal and breast cancers that prompted local fears aren't the kind normally linked to exposure to radioactivity. On the other hand, the rate of leukaemia – which is linked to radiation – is lower than average. This has led some scientists to suspect that the Gwynedd cancers are actually due to factors such as diet.

While epidemiologists now use the latest statistical methods and computer technology to test these possibilities, their aim is the same as that of the pioneers of epidemiology: to solve the mystery of ill-health among a group of people.

It was the mystery of a cancer affecting London's chimney-sweeps which attracted the attention of the eminent eighteenth-century surgeon Sir Percivall Pott of St Bartholomew's Hospital. Pott noticed that chimney-sweeps seemed especially prone to a form of skin cancer. As the cancer only seemed to affect adult sweeps, one plausible explanation was venereal disease. But Pott suspected the true cause was the constant exposure to black soot. His report on the prevalence of the disease, published in 1775, was the first to identify a cause of cancer among a specific group of people – making it the first epidemiological study. In common with many subsequent studies, it also prompted action to protect those at risk, despite the fact that the precise cause of the health risk was unknown (and in the case of soot, was not identified for another 150 years).

Even as late as the 1850s, the cause of infectious diseases was unknown, with medical scientists still clinging to such medieval notions as "miasmas".[2] Epidemiology provided crucial clues to the truth. In 1842, an official study of the condition of England's workforce by Edwin Chadwick uncovered a direct link between ill-health and poor sanitation. A Royal Commission followed, leading to the creation of a national

health board to co-ordinate basic sanitation issues from street cleansing to sewage disposal.

Around the same time, a study of the spread of measles on the Faeroe Islands between Iceland and the Shetlands led the Danish physician Peter Panum to conclude that the disease was being spread by some sort of contagious agent. It was a theory which received dramatic confirmation during an outbreak of cholera in London in 1854.

John Snow, a physician living in Soho, believed the cause to be contaminated water, having studied a previous outbreak in the south of the capital and noted that its symptoms were consistent with an ingested agent. Within 3 days of the start of the new outbreak, over 100 people in Snow's neighbourhood had died. While others fled the area, Snow stayed and gathered data in the hope of substantiating his theory.

Plotting the location of each case, Snow found that most of the deaths took place within a short distance of a water-pump on the corner of Broad Street and Cambridge Street. He took

JARGON BUSTER

Cohort study: A technique for revealing possible causes of disease in which two groups – "cohorts" – of people are identified, one of which has had exposure to a suspected risk-factor. The health of each group is then followed, and the relative rates of death or disease are compared.

Case-control study: Another method for

identifying causes of disease, in which people who already have the ailment – the "cases" – are compared with matched healthy people – "controls" – while looking at their past history to identify likely causes.

Confounder: A factor which can fool researchers into linking a disease to some apparent cause, when in

fact the link does not exist. For example, studies have shown that vegetarians are more healthy than omnivores. While this may be the result of their diet, a possible "confounder" is the fact that vegetarians taking part in such studies tend to come from higher socio-economic backgrounds and have healthier lifestyles anyway.

Descriptive epidemiology: The study of the causes of health effects using simple data analysis based on time, place and personal details.

Analytic epidemiology: The search for the cause of health-related effects using statistical analysis of the results from cohort or case-control studies.

The darker side of Snow

For tracking down the source of London's cholera epidemic in 1854, John Snow is widely regarded as one of the heroes of epidemiology. He even has a pub named after him in Soho, near the site of the infamous water pump which supplied the infected water. Yet recent historical research has raised questions about both his work and his reputation.

Studies of Snow's own data on the location and timing of deaths in Soho have revealed that the cholera epidemic was fading even before the pump was disabled. It now seems Snow's conclusions were based primarily not on evidence, but on his belief in the then unpopular germ theory of disease.

The cholera outbreak led to calls for new hygiene laws to be introduced – some of which threatened factories, whose fumes belched over cities. Astonishingly, Snow came out in defence of companies fighting the legislation, claiming that workers could not be affected by airborne pollution. He based his argument on his cholera research, insisting that all other diseases were water-borne too.

Snow's ludicrous extrapolation was bitterly criticised by other scientists. Even so, when the new laws were enacted, they were far less draconian than factory-owners had feared.

The science of epidemiology received a major boost around a century ago, with the emergence of statistical methods for extracting insight from data. In 1912 Janet Lane-Claypon at the London School of Medicine published a ground-breaking study of two groups – "cohorts" – of babies, each fed cow's milk and breast milk respectively. Lane-Claypon found that those babies fed breast milk gained more weight, and she used statistical methods to show that the difference was unlikely to occur by fluke alone. She also investigated whether something other than the type of milk could account for the difference, an effect known as "confounding".

Having demonstrated the power of cohort studies, Lane-Claypon went on to develop another key type of epidemiological investigation, the so-called "case-control" study. In 1923, as part of a government probe into the cause and treatment of cancer, Lane-Claypon tracked down hundreds of women with a history of breast cancer – the "cases" – and compared them with women who were free of the disease but otherwise broadly similar, known as "controls". This allowed Lane-Claypon to identify many risk-factors, such as numbers of children and age at menopause, still recognised as important today.

his findings to the local parish council, and persuaded them to remove the handle from the pump. Within days the outbreak ceased – leaving Snow convinced he had found the cause. He had a much harder time convincing others, however – and, it now seems, not without reason. An official report concluded Snow's data were inade-quate support for his claims about water-borne infection – and the cesspools and open sewers continued to contaminate water supplies.

Snow's pioneering epidemiological study was simple by modern standards; unfortunately, it took a simple case of cause and effect to vindicate him. During the late 1850s a massive sewer-building programme began, and within a decade uncontaminated water was available to virtually all of London. All, that is, apart from the East End, which in 1866 was the sole part of the capital struck down by cholera.

Finally, in yet another ground-breaking cohort study published in 1926, Lane-Claypon showed that rapid treatment held the key to survival among women with breast cancer. Yet despite making these huge advances, Lane-Claypon's work failed to have the impact it should – not least because she married shortly afterwards and retired from science, leaving no acolytes to carry on her brilliant work.

By the 1950s other researchers had rediscovered the cohort and case-control approaches, and were using them to probe major medical issues. In the UK, the dramatic rise in cases of lung cancer – normally a rare form of the disease – prompted Richard Doll and Austin Bradford Hill to set up a case-control study to identify the cause. By matching 1465 people with lung cancer to a similar number without the disease, Doll and Hill revealed the key factor involved: cigarette smoking.

A similar link between smoking and heart disease was uncovered in 1960 by a huge cohort study set up in 1948 by the US National Institutes of Health. Details about the health and lifestyles of over 5000 men and women living in the Massachusetts town of Framingham were collected in an attempt to uncover risk factors for heart disease. The Framingham Heart Study also quickly identified high blood pressure, obesity and cholesterol as contributors to heart disease, while showing the importance of diet and physical exercise for health.

The study continues to this day, along with a host of other epidemiological studies into the health effects of everything from coffee and mobile phone use to personality type and social status. The results are often surprising: for example, drinking a glass of wine a day has proved to be healthier than drinking none at all, reducing the risk of heart disease. Others are perplexing – such as the possibility that lack of exposure to germs in childhood increases the risk of allergies and asthma. First put forward in 1989, this "Hygiene Hypothesis" remains deeply controversial.[4]

Cigarette smoking – 50 years since Doll's discovery of the link between lung cancer and cigarettes

The publication of Doll and Hill's study linking lung cancer to smoking in 1954 began one of the biggest public health campaigns in history. Though the British Medical Council immediately declared cigarettes to be a direct cause of cancer, the first Government health warnings did not appear on cigarette packets until 1971. By the early 1980s, attention began to move to the risks to non-smokers from so-called passive smoking, with epidemiologists setting up studies to measure the health effects. A review of the results published in 1997 by researchers at St Bartholomew's Hospital, London, pointed to a significant risk of both lung cancer and heart disease from passive smoking – prompting calls for smoking bans in offices and public places.

Epidemiology continues to shape national health policies. Studies of lung cancer and heart disease rates among those living with smokers – so-called "passive smokers" – have led to worldwide moves to ban smoking in public places. Workers in countless industries now enjoy better health following epidemiological studies of the risks they face from exposure to everything from asbestos to wood-dust.

The remit of epidemiologists now spans the world, as they keep watch for unusual patterns of disease, and give warning of the emergence of deadly new infections. A puzzling increase in deaths from unusual forms of pneumonia and cancer among drug abusers and gay men in America in the late 1970s presaged the global Aids epidemic. Epidemiologists recently played a key role in stemming the spread of the Severe Acute Respiratory Syndrome (SARS) virus.

There is no cure for either of these deadly diseases, but epidemiologists have undoubtedly saved countless lives. For in public health, to be forewarned is to be forearmed.

Notes

1. Born in London in 1714, Percivall Pott rose from humble origins to become one of the most celebrated surgeons of his day. After training at St Bartholomew's Hospital, he became a full surgeon at the age of 35, and treated many famous men, including Samuel Johnson and Thomas Gainsborough. He is best known today for identifying Pott's fracture of the ankle.

2. Until around a century ago, infectious diseases were still widely regarded to be caused by foul-smelling air known as "miasma", from the Greek term for pollution. The ancient belief was not entirely groundless: bacteria-laden water or solid material often gives off an appalling smell. In June 1858, London's notorious cholera epidemic coincided with the "Great Stench", which lasted for weeks. But some Victorian scientists went further, blaming miasmas for other conditions; in 1844, one professor declared: "From inhaling the odour of beef, the butcher's wife obtains her obesity." By the end of the nineteenth century, research by Louis Pasteur and others had proved miasmas were just a by-product of the real cause: germs, in the form of bacteria and viruses.

3. One of the forgotten pioneers of modern medical science, Janet Lane-Claypon was born in 1877 into an affluent Lincolnshire family. A brilliant student at the London School of Medicine for Women, her first research was in biochemistry and hormonal control of lactation. She then moved into epidemiology, carrying out pioneering research in the causes of cancer. She died in 1967.

4. The idea that too much cleanliness may be unhealthy has captured the public imagination, but it remains deeply controversial among medical scientists. Attempts to confirm the "Hygiene Hypothesis" have produced mixed results, and uncovered some possible confounders – effects that can mislead scientists trying to find the cause of ailments. For example, research suggests that children living with pets are less prone to allergies. As pets bring in bugs and dirt, this seems to support the Hygiene Hypothesis. However, a study published by researchers at Karlstad University, Sweden, uncovered another, rather less exciting, explanation: families whose children suffer from allergies simply avoid having pets in the first place.

Further reading

www.epidemiology.ch/history/

www.framingham.com/heart/

www.ph.ucla.edu/epi/snow.html

THE PHYSICS OF REALITY

18
Special Relativity

IN A NUTSHELL

In 1887, two American scientists made the startling discovery that the speed of light always has the same value, no matter how fast observers travel relative to it. This flatly contradicted "common sense" laws of motion dating back to Galileo, and led Einstein to propose that the speed of light – and all the laws of physics – must be the same no matter how fast we move. This simple idea led to the astonishing conclusion that the properties of ordinary objects, such as their length and mass, must vary according to their speed relative to us. Known as the Special Theory of Relativity, its consequences have revolutionised science. It led directly to $E = Mc^2$, the equation that underpins nuclear power and H-bombs, and when combined with the rules of quantum mechanics, showed why chemical elements exist, and led to the prediction of anti-matter. Its ban on faster-than-light travel is now being challenged, however – opening up the prospect of Star Trek-style voyages to the stars. There are also signs that Special Relativity may not be the final word, and will have to give way to another, more sophisticated theory of space and time.

How do you start a scientific revolution? One clever experiment can be enough, or a lucky discovery – or sheer hard work. Yet there is another route, one favoured by scientists of genius. It is to ask a simple question, and to follow its consequences.

The trick, of course, is to ask the right question. In 1665, a 22-year-old student saw an apple fall at his home in Lincolnshire, and asked himself whether the same force pulled the moon towards the Earth. By pursuing his question, Isaac Newton made an intuitive leap unprecedented in scientific history, and discovered the law of universal gravitation.

In 1895 a German teenager asked what he would see if he travelled at the speed of light. The answer would lead the 16-year-old Albert Einstein to cosmic insights that ultimately eclipsed even those of the Newtonian revolution. So vast were their consequences that initially not even he was able to grasp them.

TIMELINE

Einstein's solution to his question is known as Special Relativity, and it underpins the whole of modern science. Particle physicists rely on it to understand the sub-atomic world, while astrophysicists use it to fathom the power-source of the stars. Its consequences underpin the whole of chemistry, and led to the discovery of anti-matter. It has brought nuclear generated electricity to millions of people around the world – but has put millions more under the shadow of nuclear warfare.

All this came from Einstein's realisation that something very odd would happen if he could travel alongside a beam of light. As he drew level, the beam would appear stationary – yet according to the laws of physics, a beam of light is by definition a moving electromagnetic field. So how could it ever be stationary? Did the laws of physics break down at the speed of light – or was there something else that could explain the paradox of the stationary beam of light?

Einstein was not the first to notice there was something odd about light beams. Since the late seventeenth century, many scientists thought of light as waves travelling through something known as the aether, a kind of fluid that filled all of space. If it existed, the aether would lead to changes in the measured speed of light, depending on the speed of the observer. Yet despite their best efforts, scientists could find no hard evidence for the existence of this aether.

Matters came to a head in 1879, when a new way of detecting the aether was proposed by the Scottish physicist James Clerk Maxwell. It was taken all the more seriously for coming from Maxwell, who was already famed for having revealed the cosmic unity of electricity, magnetism and light. First to take up the challenge was the American physicist Albert

Michelson, who devised an ingenious system of light beams capable of detecting the effect of the aether on light, due to the Earth's movement round the sun. His first attempt, in 1881, drew a blank. Six years later, he tried again, with an even more sophisticated experiment performed in collaboration with Edward Morley. Yet once again, they failed to find any signs of the aether: the speed of light seemed unaffected by its presence.

By now, scientists were getting desperate – and so were their explanations. In 1889, the Irish physicist George FitzGerald suggested that as it travelled through the aether the apparatus used by Michelson had shrunk by just enough to cancel out any effect. His claim was ridiculed, until the distinguished Dutch physicist Hendrik Lorentz came up with the same idea three years later, along with a formula for the shrinking effect.

Many scientists regarded it with deep suspicion, but no one could think of an alternative – no one, that is, apart from a 21-year-old physics graduate named Einstein. As a student, he had heard about the Michelson–Morley experiment, and wondered how it related to his teenage idea about light beams. He had also read works by the philosophers David Hume and Ernst Mach, which led him to reject the idea of a stationary universal fluid like the aether. To him it implied the existence of a God-like vantage point – and the only one offering the "right" view of the universe.

Einstein's rejection of the aether was also based on something he had noticed in the laboratory. Experiments with magnets and conductors gave the same results regardless of whether it was the magnet or the conductor that was moving: all that counted was their relative motion. Puzzlingly, however, Maxwell's famous laws of electromagnetism did make a distinction.

Einstein wrestled with all these strands for several years after graduating, looking for a single, coherent answer.[1] The breakthrough came in 1905, when he realised he could explain everything from two simple

JARGON BUSTER

Time dilation: The "stretching" of time observed in moving clocks, making them appear to run more slowly than they do when at rest. This has been experimentally observed in sub-atomic particles, which decay more slowly when travelling at high speed.

FitzGerald–Lorentz contraction: The shrinking of the dimensions of objects in the direction of travel which becomes apparent at very high speeds. Named after the two physicists who suggested the phenomenon to explain the Michelson–Morley experiment result, before Einstein developed relativity.

Lorentz invariance: A property of any theory if it is to be consistent with relativity, allowing its predictions to be the same – "invariant" – for all observers, regardless of how they are moving. Simple laws of motion are not Lorentz invariant.

Relativistic velocity: the speed of an object for which the effects of Special Relativity become significant. Even for a spacecraft travelling at 1 million km/hr, these are negligible, but for particles travelling at "relativistic" speeds of 90–99 per cent that of light (almost 300,000 km per second) they become crucial.

Aether: A kind of fluid once thought to fill the whole of space in order for electromagnetic waves like light to travel through a vacuum. Failed attempts to detect the aether prompted Einstein to develop the Special Theory of Relativity.

What you would see if you travelled at light speed

Special Relativity predicts that objects approaching light speed shrink in the direction of motion, and appear blindingly blue-white as they approach, and then fade to dull red-to-black as they recede. But it also predicts some distinctly counterintuitive effects. These were first noted by the Viennese physicist Anton Lampa in 1929, and were rediscovered thirty years later by Roger Penrose in England and James Terrell in the US. For example, a rocket whizzing past would appear squashed, but would also be slightly twisted around, allowing parts of it not normally visible to come into view (a phenomenon known as Penrose–Terrell Rotation). In contrast, a sphere always remains perfectly circular, no matter what its speed.

principles: that the laws of physics are the same for everyone, no matter what their speed, and that the speed of light is the same, regardless of whether its source is moving or not.

The first principle seems pretty innocuous – but as Einstein had already noticed, Maxwell's laws of electromagnetism didn't obey it, and needed modification. What kind of modification emerged from the second principle, which was far more radical. It implied that light beams do not obey the normal rules about speeds. For example, to measure the speed of a train, the speed of the observer is clearly crucial. To passengers at a station, the train's speed may be 200 km/hr, but to a car travelling alongside, the train may be doing only 100 km/hr. Which one is right? In the case of light, the answer was supposed to be the speed measured relative to the aether. Yet according to Einstein, the answer will be the same for everyone: the speed of light in vacuum, or almost 300,000 km per second.

Einstein based his radical proposal partly on Michelson and Morley's experimental evidence, but also on his teenage musings, which led him to recognise the problems of light

appearing stationary. Yet when combined with his first principle, everything fell into place – as Einstein revealed in a revolutionary paper published in 1905.

Called *On The Electrodynamics of Moving Bodies*, it sets forth what has become known as the theory of Special Relativity – so-called because it applies to the special case of bodies moving at fixed speeds relative to one another. Einstein showed that his two principles led to formulas identical to those concocted by Lorentz to explain the Michelson–Morley experiment, but without needing an "aether". The same formulas also rid Maxwell's equations of their quirks, so that they worked for all observers – and confirmed that the properties of moving objects really would appear to change as they approached the speed of light.

But the most amazing consequence of Special Relativity emerged a few months later, in a paper entitled "Does the inertia of a body depend on its energy content?" The question mark is significant, for it shows even Einstein himself could scarcely believe what he had found by combining his theory with the law of energy conservation: a formula revealing that mass was an incredibly potent source of energy. Any object with mass M could be thought of as an amount of energy E given by $E = Mc^2$, where c is the speed of light – or, put more graphically, a kilogram of matter can be thought of as energy equivalent to the detonation of over 20 million tonnes of TNT.

While Einstein wrestled with the full implications of this astounding result, scientists set about testing his other predictions. At everyday speeds, the effects of relativity are tiny: even travelling at 30,000 km/hr, the Space

Shuttle shrinks by barely 20 millionths of a millimetre. Even so, experiments with fast-moving electrons and other particles confirmed that objects really do change their characteristics in precisely the way Einstein predicted.

Theoretical physicists also began incorporating Einstein's principles into their work, with equally astounding results. In 1928, the British theorist Paul Dirac combined relativity with quantum mechanics, the laws of the sub-atomic world, and found that particles like electrons must possess a property called "spin". Named because of loose similarities with rotation, spin is now recognised as the most funda- mental property of particles, and explains why only certain arrangements of electrons are found in atoms. As such, it underpins the Periodic Table of the Elements – the basis for the whole of chemistry. Dirac also found that his theory predicted the existence of a new type of particle whose properties were like mirror-images of those of the electron, known as the positron. Four years later, studies of cosmic rays revealed hard evidence for this bizarre "anti-matter".

By the late 1930s, scientists began to wake up to the most profound impli- cation of special relativity, $E = Mc^2$. Among the first were astronomers, who recognised that this equation held the key to the power-source of the sun. In 1938, the German theorist Hans Bethe showed that the fusion of hydrogen nuclei deep within the sun led to the formation of helium "ash" weighing slightly less than the original nuclei – the difference being released as energy via $E = Mc^2$. Given the vast amount of energy thus generated by each kilogram of hydrogen and the sun's sheer size, the ability of stars to

Einstein is wrong!

Barely a year after being published, Special Relativity had been put to the test in the laboratory – and comprehensively falsified. In experiments with fast-moving electrons, Walter Kaufmann of the University of Göttingen had found results that appeared to be in better agreement with rivals to Einstein's new theory of space and time. Leading physicists of the day could not see any obvious flaw in Kaufmann's work, but Einstein's own attitude was telling. While congratulating Kaufmann on a job well done, he refused to believe the results on the entirely subjective grounds that the rival theories seemed to him less plausible than Special Relativity. He was proved right – but it took a decade before clinching evidence finally emerged.

shine for billions of years was no longer a mystery.

Back on Earth, scientists realised that $E = Mc^2$ also explained the seem- ingly endless energy pouring out of radioactive materials like uranium, and began to harness it. In 1942, a team led by the Italian physicist Enrico Fermi in Chicago created the first controlled nuclear fission reactor using uranium – demonstrating a technology that today produces around 17 per cent of the world's electricity. Three years later, the devastating effect of uncontrolled nuclear fission was demonstrated in the detonation of two atomic bombs over the Japanese cities of Hiroshima and Nagasaki.

The emergence of nuclear weapons filled Einstein with horror: "The unleashed power of the atom has changed everything save our modes of thinking," he wrote in 1946. "And we thus drift towards unparalleled catas- trophe." It is a sentiment still relevant today, with the world in fear of terrorists wielding the awesome power of $E = Mc^2$.

Even now, the implications of Special Relativity are still being explored. Its apparent ban on faster-than-light travel is under challenge from theorists using

Einstein's extremely complex extension of Special Relativity, known as General Relativity (GR). In 1994, Miguel Alcubierre at the University of Wales showed that GR allowed space and time to be "warped" in such a way as to mimic the effect of faster-than-light travel. As yet, however, no one knows how to create the required warping.

Within the last few years, cracks have begun to appear in the edifice of Special Relativity.[2] It insists that there is a limit on the energy of cosmic rays from deep space – yet ultra-fast rays breaking this limit have been detected several times in recent years. Some theorists think the answer may be that the speed of light is not alone in having the same value for all observers; there may be an energy threshold whose value is also "universal".

Work on this so-called Doubly Special Relativity is still in its infancy. Even so, it may prove to be the first glimmerings of a theory with implications even more astounding than those that emerged from Einstein's teenage day-dream.

Notes

1. In January 1990, Evan Walker, an American historian, created a sensation by claiming that relativity was not discovered by Einstein, but was really all the work of his first wife, Mileva Maric. It would be easy to reject such a claim out of hand, were it not for the fact that Mileva was a physics student with Einstein, and was quite capable of coping with the comparatively simple mathematics behind Special Relativity.

   Walker's claims were not, in fact, the first time that rumours of Mileva's role in relativity had surfaced. In 1929, a student friend had told a journalist that Mileva did not like to talk about her involvement in Einstein's famous work, for fear of casting doubt on his now worldwide reputation as a scientific genius. And in 1969, Mileva's biographer and fellow Serb Desanka Trbuhovic-Gjuric alleged that Einstein once said: "Everything that I have created and attained I owe to Mileva".

   Most Einstein experts reject such claims, pointing out that they are based primarily on hearsay and subjective interpretation of evidence. For example, Einstein's apparent "admission" of Mileva's involvement can also be seen as a touching tribute to her emotional, rather than intellectual, help in developing the theory.

   Most historians believe that her role in Einstein's career is best summed up by the testimony of her own son, Hans Albert, given to an interviewer in 1962: "Mileva helped him solve certain mathematical problems, but nobody could assist with the creative work".

2. While Einstein was hailed as a genius by many, he also had his detractors, some of whom claimed his Special Theory of Relativity was based on fundamental misconceptions. Nor were they all the autodidacts who still plague university physics departments (and science writers) with "proofs" that Einstein was a fraud. They included the Nobel Prize winners Frederick Soddy and Ernest Rutherford, and Louis Essen, the inventor of the caesium clock (which is something of an irony, as we shall see). But the most impressive critic of relativity was Professor Herbert Dingle of University College, London, and President of the Royal Astronomical Society in the early 1950s.

   Originally a supporter of Einstein's theory and an author of a textbook on the subject, Dingle came to doubt its foundations after reading an account of the so-called "clock paradox". According to this, a clock that moves relative to another will appear to run more slowly as judged by the stationary clock. Dingle claimed that Einstein's results were inconsistent with those worked out using a "commonsense" method. However, other experts showed that Dingle's method was riddled with old Newtonian concepts that Einstein had shown were invalid.

The coup de grace for Dingle's claims came in 1971, when two American scientists from the US Naval Observatory took two extremely accurate caesium clocks on flights around the world in different directions. According to Einstein, the two clocks should have registered a difference of 275 billionths of a second. The scientists measured a difference of

273 plus or minus 7 billionths of a second – exactly as Einstein's theory predicted.

Further reading

Einstein's Miraculous Year edited by John Stachel (Princeton University Press, 2005)

Subtle is the Lord by Abraham Pais (Oxford University Press, 1982)

Special Relativity: A First Encounter by Domenico Giulini (Oxford University Press, 2005)

"Seeing Relativity" Online demonstrations of the bizarre optical effects of special relativity by Dr Antony Searle and colleagues at the Australian National University:
www.anu.edu.au/Physics/Searle/
www.anu.edu.au/Physics/Searle/

19
Quantum Entanglement

IN A NUTSHELL

In the 1920s, scientists found that Quantum Theory implied that particles only have clear-cut properties when they are measured. Einstein insisted this merely showed that Quantum Theory was incomplete, and devised an experiment to prove it.

When finally performed in 1982, the experiment proved Einstein wrong – and revealed that particles can become "entangled", allowing them to correlate their properties with each other instantly, regardless of distance.

Once regarded as a delicate and esoteric effect, Quantum Entanglement is proving surprisingly robust, and is likely to be one of the key concepts of twenty-first century technology. Entangled particles are already being used to create secure communication systems; they could also be the basis of ultra-fast quantum computers, and even Star Trek-style "teleportation" machines.

Theorists now think entanglement may be relatively common in nature, raising the possibility that we live in a truly cosmic web of connections transcending space and time.

As a mere 30-something researcher with no permanent post, Alain Aspect knew he was taking a career gamble by attempting to prove Einstein wrong. Hearing of his planned experiment, one leading scientist muttered darkly: "You must be a very courageous graduate student."

Yet today, more than 20 years after he performed the experiment, Aspect is an academic celebrity, his name guaranteed a place in scientific history.[1] For he succeeded in proving that the intuition of the greatest physicist of modern times had, for once, been wrong.

What Aspect and his colleagues found in 1982 went much further, however. They uncovered the first hard evidence for a phenomenon so weird that even today most scientists struggle to grasp its implications: Quantum Entanglement.

As its name suggests, Quantum Entanglement binds sub-atomic particles together – but it does so in ways that simply boggle the mind. Perform an

TIMELINE

1923   French physicist Prince Louis de Broglie introduces the idea that particles have wave-like properties, later described as "wave functions".

1935   Einstein and colleagues put forward a "thought experiment" that they hoped would prove the incompleteness of Quantum Theory.

1935   Austrian physicist Erwin Schrodinger, whose equation governs wave-function behaviour, coins the term "entanglement" for the first time.

1964   Irish physicist John Bell proves mathematically that the existence of entanglement can be demonstrated and measured in the laboratory.

1982   Alain Aspect and colleagues at the University of Paris demonstrate entanglement for the first time, proving Einstein wrong about quantum effects.

1985   David Deutsch at Oxford University shows that entanglement could open the way to "quantum computers", which solve a myriad problems in parallel.

1993   International team led by Charles Bennett of IBM show that entanglement allows Star Trek-style "teleportation" – at least of sub-atomic particles.

1997   Prof. Anton Zeilinger and colleagues at the University of Vienna use entanglement to "teleport" a photon 1 metre across their laboratory.

2001   Team led by Eugene Polzik at the University of Aarhus, Denmark, successfully entangle two clouds of trillions of cesium atoms.

2003   Prof. Anton Zeilinger and colleagues successfully transmit entangled photons through the air across the river Danube.

2004   Theorist Caslav Brukner and colleagues at Imperial College, London, show that Quantum Theory may allow entanglement over time as well as space.

prospect of distinctly futuristic applications, from ultra-fast computing to Star Trek-style "teleportation", beaming objects across the cosmos.

Perhaps most astonishing of all, evidence is now emerging that entanglement is widespread throughout the universe, hinting at a level of cosmic connectedness of truly New Age proportions.

The nature of this connectedness is far from intuitive, defying the usual bounds of space, time and common sense. It can only be fully understood through the rules of Quantum Theory, and in particular a series of discoveries that began with an idea suggested by a French aristocrat in 1924.

In his doctoral thesis, Prince Louis de Broglie pointed out that recent experiments had revealed that light, usually regarded as a wave, sometimes behaved as if it came in particle-like packets dubbed photons. So perhaps, he argued, particles like electrons might sometimes behave like waves, whose properties he estimated using simple arguments.

It was a daring prediction, yet just 3 years later it had been confirmed in experiments that showed that supposedly solid electrons could generate the wave-like effect known as interference. Many leading physicists, including Einstein, saw this as a major breakthrough, and set about trying to find out more about these strange "matter waves".

Among them was Erwin Schrodinger, an Austrian theorist who constructed an equation that governed their behaviour. This revealed that the waves aren't like any ordinary, physical waves at all, prompting Schrodinger to give them an altogether more abstract name: "wave functions".

Schrodinger's discovery was a blow to those who – including Einstein –

action on one particle, and its entangled partners are affected instantly, no matter where they are, or how far apart.

It sounds like something out of science fiction, and it opens up the had hoped Quantum Theory would give them a clear, common sense view of reality. Worse was to come, with the German theorist Werner Heisenberg uncovering the famous uncertainty principle, which showed it was impossible to know everything about sub-atomic particles at any given time.

By the 1930s, the sub-atomic realm had emerged as an utterly bizarre place, populated by fuzzy particles with no definite properties until jolted by the act of observation. For Einstein, it all just meant that something was missing from Quantum Theory. And in 1935, together with Boris Podolsky and Nathan Rosen, he proposed an experiment to prove it.

Known as the "EPR experiment", it was brilliantly simple.[2] Picture a single molecule that explodes into two iden-tical particles, which fly off in opposite directions. According to Quantum Theory, neither particle has clear-cut properties until it is observed. So suppose we want to know the position of one of the particles: we observe it, the resulting "jolt" supposedly forcing it to acquire definite properties.[2]

Then Einstein sprang his trap: once we know the position of the first particle, we can use Newton's laws to figure out the position of its partner, without ever having to observe it. There's no uncer-tainty involved – and no need for the supposedly crucial "jolt" of observation to ensure it acquires definite properties. In short, said Einstein, the particles always have definite properties, whether they're observed or not. The fact that Quantum Theory can't account for this, he declared, merely showed it was incomplete.

There was a loophole, however. Perhaps the first particle somehow signalled the fact it had been observed, prompting its still-fuzzy partner to take on a definite state. But Einstein had a knock-out argument against this, too: his own theory of relativity, which showed no signal can travel faster than light. This rules out any hope of the particles instantly communicating with

JARGON BUSTER

Spin states: Many particles, including the electrons, protons and neutrons in atoms, possess a quantum property known as "spin", which is loosely akin to rotation. The major difference is that it can only take particular values, and if the particles are created in the right way, these so-called spin states can become entangled.

Wave-function: A math-ematical quantity which captures the key prop-erties of a particle, or set of particles. Individual particles have their own wave-functions, but if they are prepared in the right way, two or more particles can have wave-functions that become entangled with one another, so they share the same wave-function.

Bell's Theorem: A mathe-matical result found by the Irish physicist John Bell in 1964, which shows that entanglement produces a greater degree of correlation between separated particles than can be accounted for by common sense.

Spontaneous Parametric Down-Conversion (SPDC): A standard method of triggering entanglement, in which some of the photons of light shone into certain crystals split in two, creating pairs of entangled photons.

EPR experiment: An imaginary experiment named after the initials of its inventors – Einstein, Boris Podolsky and Nathan Rosen – in which a single particle splits in two, creating entangled pairs. Put forward in 1935, the EPR experiment was finally performed in 1982.

John Bell

One of the most original theoretical physicists of the twentieth century, John Bell was born in Belfast in 1925, and entered Queen's University at the precocious age of 16. He quickly focused his attention on the quantum world, first as a designer for particle accelerators, and then as a theorist. His most famous paper stemmed from his life-long belief that Einstein was right in his suspicions about Quantum Theory, while Neils Bohr talked only in riddles. Bell set about capturing the argument over quantum theory with such clarity that it could be put to the test. Despite his work being widely held to have proved Einstein wrong, Bell himself insisted that Aspect's findings were not as clear-cut as many believed.

each other. They simply had to have definite properties all the time.

Einstein's argument was attacked by defenders of Quantum Theory, but without hard experimental data neither side could deliver a knock-out blow. The prospect of resolving the debate only emerged 9 years after Einstein's death in 1955.

The Irish theorist John Bell proved that if particles have permanent properties as Einstein claimed, then there is a limit to how similar the properties of the pairs of particles in his experiment could be, set by the finite speed of light. If Einstein was right, the correlation in the properties of the particles could never exceed this "Bell Inequality".

In 1982, Alain Aspect and his colleagues at the University of Paris finally succeeded in performing the EPR experiment using photons of light. And they found the Bell Inequality was utterly broken: astonishingly, even pairs of photons separated by large distances seemed able to "communicate" with one another instantly, giving a level of correlation far exceeding the Bell Inequality.

The implications were dramatic. As the Bell Inequality had been broken, at least one of its assumptions – and thus those made by Einstein – must be wrong. Either particles do not have definite properties until they are observed, or they are able to communicate faster than light – or even both.

Yet for many physicists, the take-home message was more mundane: Quantum Theory had been proved correct – which, given its enormous successes, was hardly a surprise. Nor were they worried about the idea of faster-than-light signalling. As it's impossible to control the outcome of measuring the properties of one entangled particle, it's also impossible to dictate what it sends to its entangled partner, making it useless as a means of communication.

Aspect's experiment thus seemed to many to be an esoteric tidying-up exercise, involving a subtle phenomenon of no practical value. It has taken physicists some time to get to grips with the awesome implications of Quantum Entanglement. Now they have, and the results are proving to be truly mind-boggling.

Among the first was theorist David Deutsch at Oxford University, who in 1985 showed that entangled particles could be exploited in a "quantum computer", which solves problems far faster than any current machine.[3] Put simply, while a conventional computer uses electrons whose properties are well-defined, a quantum computer uses the fact that entangled particles possess a range of possible states, each of which can be used to carry out binary calculations. This allows lots of problems to be solved in parallel, making such a computer unimaginably faster than today's supercomputers.

So far, creating the necessary hardware to build a quantum computer has proved difficult, but many physicists believe it can be done.

Their confidence comes from the fact that entanglement has proved much more robust than originally thought. In 2002, a team at Leiden University in the Netherlands showed that the entanglement effect can pass through metal, and one cold, windy night in 2003, Anton Zeilinger and colleagues at the University of Vienna successfully beamed entangled photons through the night sky across the Danube.[4]

The first real-life applications of entanglement are also starting to emerge. In 1990, Artur Ekert at the University of Oxford suggested that entanglement could be the basis of an unbreakable code system. It involves digitising messages and adding random "key streams" to them, garbling the result; only a person with the right random key can reverse the process and read the message. While provably unbreakable, the big problem is ensuring the key reaches only the right people. Ekert showed that entangled photons could do the trick, as any attempt to intercept them en route would produce tell-tale changes in their correct entangled states. In April 2004, Zeilinger and his colleagues successfully used this "quantum key distribution" method to encrypt a financial transfer of 3000 euros between Vienna City Hall and Bank Austria Creditanstalt.

But most astonishing of all is the prospect of using entanglement to "beam" objects from one place to another, like the Star Trek transporter. First suggested by theorists in 1993, it involves scanning the object and using pairs of entangled particles to capture its properties. Information about the object captured by one set of entangled particles can then be used to recreate the object wherever their entangled partners have been sent.

How entanglement is tied to knots

The very word "entanglement" conjures up images of knotted-up threads, but it seems obvious that there can't be any real connection between the abstract "wave functions" of Quantum Theory and the world of granny-knots and shoe-laces.

Yet, amazingly enough, just such a connection has been found by physicist Padmanabhan Aravind at Worcester Polytechnic Institute in Massachusetts. In 1997, he showed that the entanglement of two or more particles followed mathematical rules obeyed by basic knots and braids, with the act of observation being equivalent to cutting the knots. For example, the behaviour of three entangled particles bears curious similarities to the so-called Borromean Rings, in which cutting any one free releases the other two.

In 1997 Zeilinger and colleagues succeeded in "teleporting" a single photon over a metre using entanglement.[5] Four years later, a team at the University of Aarhus, Denmark, led by Eugene Polzik succeeded in preparing trillions of atoms in an entangled state fit for teleportation.

No-one expects to do the same with human beings any time soon; it's been estimated that the information needed would need a set of CDs filling a box with sides of 1000 km. Yet the very fact that such possibilities are being discussed by reputable scientists shows how Quantum Entanglement is revolutionising physics.

It is also changing perceptions of the very nature of the universe. Physicists have discovered that the cosmos is filled with a seething mêlée of so-called "vacuum particles", constantly popping in and out of existence. These particles are highly entangled – and the latest astronomical data suggest their effects control the fate of the universe.

The picture now emerging is one of a cosmos filled with entangled matter and energy, communing together instantaneously, regardless of distance. Very recently theorist Caslav

Brukner and colleagues at Imperial College, London, have shown that entanglement may even transcend time, linking together the past and future.

Einstein may be best known for being right, but the consequences of his failure to undermine Quantum Theory may prove to be his most astounding scientific legacy.

Notes

1. Born in 1947, Aspect became fascinated by Quantum Theory while doing voluntary work in Cameroon in 1971 after graduating in science. Now at the University of Paris-Sud, his demonstration of Quantum Entanglement led to his election to the prestigious Académie des Sciences in 2001.

2. The original EPR experiment involved an imaginary particle that exploded into two identical fragments that flew off in opposite directions. The first attempt to perform the experiment in real life was made by researchers at the University of California in 1972, using entangled pairs of photons emitted by calcium atoms. While confirming the predictions of Quantum Theory, the conclusions were only made completely water-tight by Aspect and his colleagues at the University of Paris-Sud in 1982, who ensured measurements of the photons were made only after they were too far apart to "compare notes" at the speed of light.

3. Quantum Entanglement is expected to play a key role in quantum computing, in which information is captured via so-called qubits – particles which exist in mixtures of states, rather than just the usual "binary" choice of ones and zeros. Entanglement allows the properties of each qubit to be shared among many others instantaneously, producing a dramatic increase in speed.

4. Despite its sub-atomic origins, Quantum Entanglement is proving to be surprisingly robust – as Anton Zeilinger and colleagues of the University of Austria demonstrated by beaming entangled photons to two different receivers separated by 600 metres on either side of the river Danube. The photons were created by injecting laser beams into special crystals, which split some of the photons into entangled pairs. Despite the poor conditions, with wind gusting up to 50 km/hr, the photons arriving at the two receivers remained entangled. The team found that the reception quality was equivalent to beaming the photons down from a satellite orbiting 600 km above the Earth – suggesting Quantum Entanglement could one day span the world.

5. Teleportation seems like pure science fiction: after all, it's not a copy of the object that is sent, but the thing itself. Yet that is precisely what Quantum Entanglement makes possible – and scientists have recently done it with whole atoms. Sending humans is still some way off, not least because we consist of around a billion billion billion atoms.

Further reading

Speakable and unspeakable in quantum mechanics by John Bell (Cambridge UP, 1989)

Quantum: A guide for the perplexed by Jim Al-Khalili (Weidenfeld & Nicolson, 2003)

The New Physics, edited by Paul Davies (Cambridge UP, 1988)

Entanglement by Amir Aczel (Wiley, 2003)

http://www.quantum.univie.ac.at/

http://plato.stanford.edu/entries/qt-entangle/

20
The Standard Model

IN A NUTSHELL

Since the 1920s, scientists have been building up a detailed description of sub-atomic particles and the forces between them. In the process, they have found that the properties of many particles can be explained using even more fundamental entities called quarks, while the forces that act on matter are the result of the exchange of special "exchange particles" like photons and gluons. Taken together, the result is the Standard Model, and it has led to a host of successful predictions about the nature of matter over the last 40 years. Even so, physicists accept that it cannot be the ultimate Theory of Everything – not least because it does not include gravity among its fundamental forces. Evidence is now emerging for the existence of effects beyond the reach of the Standard Model, and many physicists believe we are on the brink of a revolution in understanding the cosmos and its contents.

The task facing Carlo Rubbia and his team was mind-boggling – but then, so was the reward. They were searching for the equivalent of a few unusual grains of sugar lost among three-quarters of a tonne of the stuff. But if they succeeded, they would be hailed as scientific heroes – and be guaranteed Nobel Prizes.

Over the course of a few months in 1982, Rubbia and his team had smashed together billions of sub-atomic particles in the huge Super Proton Synchrotron accelerator at CERN, Europe's nuclear research centre near Geneva. Now they had to sift through the wreckage of those collisions, looking for tell-tale signs of two sub-atomic particles never before seen.

Called W and Z, their existence had been predicted years earlier by theorists trying to create what had defeated even Einstein: a single unified account of the forces that rule the cosmos.

The W and Z particles[1] were predicted to be incredibly unstable, decaying into other particles almost the instant they were created. Even so, calculations suggested that if a billion protons and their anti-matter counterparts were smashed together, there might be a few Ws and Zs in the debris.

At first sight, confirming this seemed impossible: there simply wasn't enough computing power to sift through so many collisions fast enough. So the team devised an electronic "sieve" which

TIMELINE

1928 British theorist Paul Dirac combines relativity with laws of electromagnetism to create Quantum Electrodynamics (QED), explaining existence of spin.

1935 Japanese theorist Hideki Yukawa puts forward an explanation for fundamental forces based on an idea of "exchange particles" transmitted from place to place.

1964 Murray Gell-Mann and George Zweig independently suggest that properties of many particles can be explained using combinations of "quarks" within them.

1964 Peter Higgs of the University of Edinburgh puts forward a means of giving mass to subatomic particles, which becomes known as the Higgs particle.

1967 Steven Weinberg and Abdus Salam use Higgs particle idea to create a viable theory that unifies the electromagnetic and weak nuclear forces.

1971 Gerardus 't Hooft uses a computer to prove that electroweak theory is free of mathematical problems that dogged attempts at unification.

1973 Sheldon Glashow and Howard Georgi put forward the first attempt to link the electroweak and strong forces in a "Grand Unified Theory" (GUT).

1983 International team led by Carlo Rubbia at the SPS accelerator at CERN, Geneva find first evidence for the W and Z particles predicted by electroweak theory.

1998 Neutrinos shown to have mass in experiments at the SuperKamiokande experiment in Japan – the first proof of effects beyond the Standard Model.

2000 The "tau neutrino", the final particle needed to complete the Standard Model, is found by an international team at Fermilab in Chicago.

2004 Evidence for breakdown of Standard Model continues to accumulate from experiments, such as studies of B-mesons at the KEKB accelerator in Japan.

instantly rejected obviously unpromising events. This eliminated 99.9 per cent of the raw data – but still left a million collisions needing scrutiny. Using sophisticated computer imaging methods, the CERN team began winnowing out all but the very best data for final analysis. They eventually ended up with just five collisions – but it was enough: the tell-tale signs of the W were there. A few months later, they also found the Z, and in 1984 Rubbia and Simon van der Meer, the engineer who had built the accelerator, got their reward: the Nobel Prize for physics.

The real prize was even bigger, however. The W and Z had exactly the properties predicted by theorists – confirming their confidence in a grand theory with a misleadingly down-beat name: the Standard Model.

Built up over the last 70 years, the Standard Model is the most successful attempt yet to create a single theory describing the forces and particles that make up our universe. It explains phenomena ranging from radioactivity to the power-source of the sun, and has sprung a host of surprises, from the existence of anti-matter to the existence of particles trapped within others. Its detailed predictions have been confirmed to a few parts in 100 million.

Not surprisingly, the Standard Model has long been regarded as one of the crown jewels of science. But now flaws are starting to appear – and they hint at the existence of an even grander theory whose power can only be guessed at.

Almost inevitably, the origins of the Standard Model lie in the work of a Swiss patent clerk named Albert Einstein. In 1905, he published a series of papers which overturned long-standing ideas about the nature of space, time and matter. His theory of

Special Relativity predicted that strange new phenomena would start to emerge at speeds close to that of light. He also showed that light itself could be thought of as a stream of energy-packed bundles called quanta.

These revolutionary insights clearly had major implications for electrodynamics, the laws governing electromagnetic radiation such as light. In 1928, a brilliant young British physicist named Paul Dirac unveiled the result of integrating these laws with Special Relativity and Quantum Theory. Known as Quantum Electrodynamics (QED), its equations turned out to be a treasure-trove of insights. A new view of the electromagnetic force emerged, in which light quanta – "photons" – flitted between charged particles, carrying the force between them. But most amazing of all, QED predicted the existence of "anti-matter", the exact opposite of ordinary matter.[2]

The discovery in 1932 of the positron – the anti-matter equivalent of the electron – in studies of cosmic rays confirmed the power of QED, and

Quantum Electrodynamics

The result of combining the classical laws of electromagnetism with Quantum Theory and relativity, Quantum Electrodynamics (QED) is one of the most successful theories ever devised. It led to the prediction of anti-matter, and gave an explanation for the otherwise puzzling quantum property known as spin – the most fundamental property possessed by sub-atomic particles. In the case of the electron, this led to predictions for the strength of its magnetic field which has been confirmed by experiment to an accuracy of 1 part in 10 billion. One of its most impressive consequences, confirmed in the 1940s, was the existence of energy emerging literally out of nowhere. So-called "vacuum fluctuations" now play a key role in fundamental physics, including theories for the origin and nature of the universe.

prompted theorists to push the same ideas further. Among them was the Japanese theorist Hideki Yukawa, who wondered if other forces of nature could also be explained in terms of "exchange particles" like photons. In 1935, he suggested that the so-called strong nuclear force, which binds together protons and neutrons in the nucleus, was transmitted via an exchange particle called a meson. He predicted its likely mass, and in 1947 the particle duly turned up in studies of cosmic rays.

JARGON BUSTER

Quarks: Trapped within all particles that feel the strong nuclear force – such as protons and neutrons – quarks come in six "flavors", different combinations of which give their host particles their observed properties. Quarks are prevented from escaping by particles known as gluons, whose grip increases as quarks try to break away.

Spin: A property of sub-atomic particles with some resemblance to the spin of a ball, although like all quantum theoretic ideas the analogy is far from perfect. All force-carrying particles like photons are "bosons" with whole-number units of spin (0,1, 2 etc.), while matter particles like electrons are "fermions", with half-integer values (1/2, 3/2 etc).

Neutrinos: Ghostly particles whose existence was predicted on theoretical grounds in 1931 and which were discovered in 1959. Barely interacting with matter, neutrinos have no charge and according to the Standard Model have no mass. In 1998, scientists announced evidence of a mass around one-millionth that of the electron – the

first evidence for effects beyond the Standard Model.

Electroweak theory: The first successful unification of two fundamental forces of nature: electromagnetism and the weak nuclear interaction. Developed during the 1960s, it is one of the most spectacular successes of the Standard Model.

Murray Gell-Mann

One of the most creative theorists alive today, the American Murray Gell-Mann (born 1929) won the 1969 Nobel Prize for physics for his contributions to bringing order to the apparently chaotic world of sub-atomic particles. By classifying particles according to certain combinations of properties, Gell-Mann discovered the existence of regularities that allowed him to predict the existence of new particles with astonishing accuracy. He went on to propose that these patterns were ultimately due to the existence of even more fundamental particles than protons and neutrons, known as quarks. More recently he has focused on attempts to create a Quantum Theory of gravity, and to find underlying rules in the complex behaviour of natural phenomena from earthquakes to forest fires.

Yukawa's particle then took its place among a list of sub-atomic particles that seemed to be growing without limit. By the late 1950s, there were dozens of them, their disparate properties defying belief in the essential unity of nature. But in 1960, theorists made a bizarre discovery. If the properties of particles were plotted on graph paper, striking hexagonal and triangular patterns emerged. Or, rather, almost emerged: there were gaps, hinting at the existence of undiscovered particles. Over the next few years, the missing particles were found – and they had exactly the properties predicted by the geometrical patterns.

But why did the patterns exist? In 1964, the American theorist Murray Gell-Mann put forward an astonishing idea: that locked inside many of the particles were truly fundamental entities he called "quarks" (rhyming with "forks").[3] The disparate properties of particles, Gell-Mann claimed, simply reflected different combinations of two or three quarks within them. It was a daring idea – not least because it implied the quarks had some very odd properties never before seen in nature.

In 1968, physicists began searching for quarks by blasting particles through protons and neutrons, probing their interiors. And the results revealed the presence of nugget-like objects inside the particles, with precisely the properties predicted for quarks.

Their discovery revived hopes for a unified theory of sub-atomic particles. Around the same time, new hope had also been injected into the quest for a unified theory of fundamental forces. As early as 1938, Yukawa had found hints that two of these forces – electromagnetism and the so-called weak force – were really just different facets of a single "electroweak" force. Yet attempts to prove the connection ran into a big problem: the short range of the force meant the exchange particles involved had to be relatively heavy, but the theories only produced massless particles like photons.

In 1964, the British theorist Peter Higgs put forward a solution, in the form of a particle which was "consumed" by others, giving them mass.[4] Now known as the Higgs particle, theorists seized on the idea, and by 1968 Steven Weinberg in America and Abdus Salam in England had independently devised a unified theory of the electroweak force, featuring heavy exchange particles dubbed W and Z.

But was it right? Confidence soared in 1971, when computer calculations showed the theory was free of mathematical problems that dogged previous attempts. But the real proof only came in 1983, when Carlo Rubbia and his colleagues at CERN finally discovered the W and Z.

By then theorists had begun constructing the so-called Grand Unified Theory (GUT), which brought the strong nuclear force into the fold as well. This forged a direct link to the emerging unified theories of particles based on quarks. Experiments had shown that these were trapped inside their hosts by "gluons" – the exchange particles of the strong force. Meanwhile, theorists found fresh connections between the properties of quarks, electrons and other particles which hinted at yet more undiscovered members of the sub-atomic family.

When the last member of this menagerie – the so-called "tau neutrino" – was discovered in July 2000, the Standard Model reached its zenith. By then it was already clear that it could not be the last word. Most obviously, the theory left out the most important force of all: gravity. Then there was the disquieting fact that the success of the Standard Model rested on a huge number of assumptions, including nineteen "free parameters", whose values defied explanation. But most worrying of all, the Higgs particle – crucial for explaining why particles had mass – had still not been found.

Any lingering complacency was shattered in 1998, when scientists at the SuperKamiokande experiment in Japan announced that the ghost-like neutrinos have mass. Although barely one-millionth the mass of the electron, it was still enough to put it beyond explanation by the Standard Model.[5]

Another major crack has appeared very recently. One of the triumphs of the Standard Model is its ability to predict the correct value of a certain magnetic property of particles called muons to better than one part per million. Now extremely precise measurements carried out by an international team at the US Brookhaven

Grand Unified Theories

During the 1960s, theoretical physicists showed that two of the four fundamental forces of the cosmos – electromagnetism and the weak force, responsible for radioactivity – were in fact different aspects of a single "electroweak" force. By the early 1970s, attempts began to unify this with another force: the strong interaction, which holds together the atomic nucleus. The result was given the somewhat inflated title of Grand Unified Theory (GUT) – inflated in that the early attempts are now known to be incorrect, and also because the theory completely fails to incorporate the fourth and most familiar force, gravity. While some progress has been made in creating a GUT, serious problems still remain. One of the most serious is that such theories predict that the proton – a fundamental building-block of atoms – is actually unstable, and decays into other particles on a timescale of around a million million billion years. So far, however, all attempts to confirm this prediction have failed.

National Laboratory have revealed discrepancies with the predictions of the Standard Model.

Again, although tiny, these discrepancies cause huge problems for the Standard Model. They also hint at the existence of new effects with major implications for our understanding of the universe. Among them is "supersymmetry", a new way of classifying particles which bridge the gulf between particles which form matter, like protons and electrons, and exchange particles like photons and gluons.[6]

Within the next few years, scientists hope to prise more secrets from the heart of matter using the gigantic Large Hadron Collider (LHC) machine, now being completed near Geneva.[7] Top of their most wanted list is the Higgs particle, whose existence is needed by the Standard Model to explain why particles have mass.

If the LHC does find the Higgs particle, its prediction will be the crowning achievement of the Standard Model. Yet, ironically, the LHC also looks set to write its epitaph, by confirming the reality of effects beyond its reach, such as supersymmetry.

Whatever the LHC reveals, the Standard Model will take its place alongside the likes of Newton's law of gravity – as a conception of the universe which, while flawed, remains one of the towering achievements of human intellect.

Notes

1. Predicted by theorists in the 1960s, the W and Z are the "exchange particles" for the so-called weak force, responsible for radioactive decay. Because of the short range of the force, the particles were expected to be relatively heavy – a prediction confirmed in 1983.

2. Predicted by the British physicist Paul Dirac in 1931, anti-matter is the exact mirror-image of ordinary matter, with the same mass but opposite charge. When the two types meet, the result is the most violent form of energy release in the cosmos, around 10,000 times more powerful than an H- bomb, kilogram for kilogram. Scientists believe equal amounts of matter and antimatter were created in the Big Bang, with effects predicted by the Standard Model leading to today's preponderance of ordinary matter.

3. In the early 1960s a graduate student named George Zweig began trying to make sense of the disparate properties of sub-atomic particles. At the time, over two dozen were thought to exist, but Zweig found he could explain all their properties – if they contained even more fundamental particles, which he called "Aces". Zweig's daring proposal was ridiculed, and he was forced to abandon attempts to publish it. Meanwhile, the brilliant – and better-known – US theorist Murray Gell-Mann came up with the same idea, and convinced others to take his "quarks" seriously. They are now at the very heart of modern physics.

4. As the eponymous theorist behind the Higgs particle, many physicists believe Peter Higgs will win a Nobel Prize – if his famous particle is discovered. During the 1960s, theorists trying to unify the electromagnetic and weak interactions found that their theories led to particles without mass. Peter Higgs put forward the idea of a heavy particle which was "eaten" by others, giving them this crucial property.

5. Thousands of metres below Mt Ikenoyama in central Japan lies the joint US-Japanese SuperKamiokande experiment, the world's most sensitive detector of neutrinos. Created when cosmic rays from space bombard the earth's upper atmosphere, neutrinos barely interact with the Earth. To detect their presence, scientists at SuperKamiokande use a tank containing 50,000 tonnes of highly purified water, and look for faint flashes of light given off when neutrinos interact with the molecules in the tank. After 2 years of work, the team was rewarded in 1998 with the discovery of evidence that neutrinos have mass – the first-ever result hinting clearly at physics inexplicable by the Standard Model.

6. While today's sub-atomic particles possess a host of different properties, theorists believe that in the very early history of the universe they all exhibited a basic similarity known as "supersymmetry". Testing this prediction is one of the aims of the Large Hadron Collider (LHC), in which scientists will accelerate particles to energies that mimic conditions very close to the Big Bang, to see if the tell-tale signs of supersymmetry start to appear.

. In 2007 particle physicists will finally have access to the machine needed to see what lies beyond the Standard Model: the Large Hadron Collider (LHC). A colossal 27 km long ring of tubes and magnets buried 100 metres below the outskirts of Geneva, Switzerland, the LHC will smash protons together with a violence never before seen on earth. The energy of the protons will be equivalent to a temperature of 100 million billion degrees C – similar to conditions just one million-millionth of a second after the cosmic Big Bang.

To keep the fast-moving protons on track, the LHC will use over 1000 giant superconducting magnets, each over 100,000 times stronger than the Earth's magnetic field. Once up to speed, the particles will be brought together in detectors, creating a huge spray of fragments. Scientists hope that within the debris they will find the answers to such long-standing mysteries as the origin of mass, which

is thought to be linked to the existence of the yet undetected Higgs particle. But they will have to be quick: out of the 800 million events that will take place inside the LHC each second, calculations suggest only one might produce anything interesting.

Further reading

A Unified Grand Tour of Theoretical Physics by Ian Lawrie (Adam Hilger, 1990)

The New Physics edited by Paul Davies (Cambridge, 1989)

The Particle Explosion by Frank Close et al. (Oxford, 1994)

http://particleadventure.org/particleadventure/

http://www2.slac.stanford.edu/vvc/theory/model.html

www.cern.ch

21
The Theory of Everything

IN A NUTSHELL

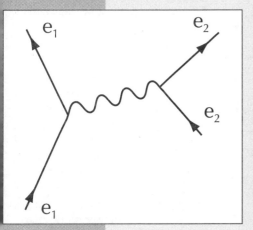

For the last 150 years, physicists have been on a quest to find the Theory of Everything, a single set of equations that describe all the forces and matter in the universe. The first success came in the 1860s, when the Scottish physicist James Clerk Maxwell showed that electricity and magnetism were actually different facets of one single phenomenon: "electro-magnetism". Now the goal is to find a way of connecting this to the other forces of the universe, and the particles on which they act. For decades, the force of gravity refused to fall into line. During the 1970s, theorists discovered that the impasse could be broken if particles were not mere points, but were more like tiny multi-dimensional "strings". For the next 20 years, these bizarre entities were the centre of attention in the quest for the Theory of Everything. Now superstrings are thought to be just a small part of something even stranger and bigger: so-called "M-theory". According to this, our universe and everything in it can be described in terms of 11-dimensional objects, with all but 4 of the dimensions curled up far smaller than an atomic universe. Although the final verdict is not yet in, M-theory is widely regarded as the closest theorists have yet come to the ultimate Theory of Everything.

On Sunday 17 April 1955, Albert Einstein sat up in his bed in Princeton Hospital, and began the last calculation of his life.

In his characteristically neat handwriting, he set down line after line of symbols. After sorting out some algebra and fractions, he put his work to one side and rested. A few hours later, the greatest scientist of the century was dead. By his bedside lay his his final, failed attempt at creating his dream of a "unified field theory": a single, coherent explanation of all the known forces in the cosmos.

Einstein had sought it for over 30 years, without success. Today, half a century after his death, his dream may be about to become a reality. Some of the world's most brilliant theoretical

TIMELINE

Its grand title and even grander aims belie its origins – in experiments performed in a Victorian laboratory during the summer of 1831. At London's Royal Institution, the great English physicist Michael Faraday was studying the relationship between electricity and magnetism. He already knew that electricity flowing through a wire produces a magnetic field; what Faraday wanted to know was whether the reverse was also true: could magnetism generate electricity?

After some false starts, he succeeded – and built the first-ever version of what we now call a dynamo. In the process, Faraday had uncovered something profound: that despite appearances, electricity and magnetism were really just different facets of the same basic phenomenon. It was the first hint of what was to become one of the guiding principles for those seeking the ultimate Theory of Everything: that there is a basic unity to the cosmos – if it is looked at in the right way.

While Faraday's experimental skills had allowed him to glimpse this unity, he lacked the key intellectual tool needed to reveal it in all its glory: mathematics. In 1861, the Scottish theorist James Clerk Maxwell succeeded in translating Faraday's discoveries into mathematical language. The result was the now celebrated Maxwell's Equations of Electromagnetism, which made the essential unity of electricity and magnetism explicit.

It was a brilliant achievement, but also one that begged an obvious question: did this cosmic unity extend to encompass that most familiar of forces, gravity? This was the challenge Albert Einstein set himself shortly after publishing his radically new vision of

physicists think they have glimpsed a majestic theory encompassing far more than even Einstein thought possible. In its final form, it will explain not only all the forces at work in the universe and the particles on which they act, but even the nature of space and time itself.

Small wonder that it's been dubbed the Theory of Everything.

gravity, known as General Relativity (GR), in 1915.

More than any physicist before him, Einstein had both the belief and intellect needed to unify gravity and electromagnetism. Yet he soon found that the challenge was far harder than he imagined. The first major stumbling-block was finding a way of capturing GR, his own theory of gravity, and Maxwell's equations in a single, unified format. According to GR, gravity is the result of warping of the very fabric of space and time around us; Maxwell's equations, in contrast, viewed electromagnetism as a kind of "force field" flowing through this four-dimensional arena.

In 1919 Einstein learned of what he thought was a big clue to the unification of these two disparate theories. Theodor Kaluza, a German mathematician, had shown that one set of equations could sum up both – but only if the universe contained an extra, fifth, dimension.

It was a stunning idea, but was it anything more than a curiosity? Where was this extra dimension? In 1926, the Swedish mathematician Oskar Klein suggested an answer: perhaps it was curled up too small to be detectable, just as the tiny width of a three-dimensional hair makes it seem only two-dimensional.

While Einstein's intuition convinced him of its importance, he could not turn Kaluza and Klein's discoveries into the unified theory of gravity and electromagnetism he sought: there were always loose ends, or ludicrous consequences. He tried other approaches, but fared no better. By the time of his death in 1955, most physicists were convinced Einstein had simply wasted his time on ingenious but empty mathematics.

It was a belief seemingly confirmed by the fact that while Einstein was grappling with his unified field theory, two further fundamental forces had

JARGON BUSTER

Planck length: the scale on which the usual notions of space and time break down under quantum effects. Space may appear to be smooth and continuous, but theory shows that if we could observe down to scales of 10 to the power minus 35 metres (so small that atoms would appear the size of galaxy clusters), we would observe space to be more like a storm-tossed ocean.

Supersymmetry: A mathematical link between the two basic types of sub-atomic particles: fermions (such as the electron), from which all matter is made, and bosons (such as the photon), which transmit fundamental forces. The discovery in the early 1970s of this unexpected link has played a key role in the development of a unified Theory of Everything.

Kaluza–Klein theories: Theories which unify the forces and particles of nature using more than the usual three dimensions of space. Theorists currently believe that the true Theory of Everything will be a Kaluza–Klein theory with 11 dimensions, all but 4 of which are too small to observe.

String theory: a description of particles and the forces between them based on the notion that particles are not mere points, but are more like tiny 10-dimensional "strings" whose size is around that of the Planck length. When combined with the idea of supersymmetry, the resulting entities are known as "superstrings".

M-theory: The current front-runner for the title of Theory of Everything. First brought to prominence in 1995, M-theory seems to be the over-arching theory of which superstrings are just a part. The M is usually taken to stand for "membrane", of which 10-dimensional superstrings can be thought of as forming merely an "edge".

Symmetry: where beauty meets truth

Symmetry has long been regarded as a key aspect of beauty, but during the twentieth century theorists found that it has deep links to cosmic truth as well. In physics, symmetries usually take the form of mathematical operations which leave quantities unchanged – just as, say, rotating a symmetric crystal through certain angles leaves it looking unchanged. Sometimes these mathematical symmetries reveal patterns in the properties of particles, gaps within which allow physicists to make predictions about as yet undiscovered particles. In other cases, the symmetries manifest themselves as conservation laws; for example, the law of energy conservation is related to a special symmetry of space-time.

been found: the so-called strong interaction, which glues together atomic nuclei, and the weak nuclear force, responsible for radioactivity. Worse still, these forces seemed best explained in terms of "messenger" particles transmitting the forces from place to place – utterly unlike Einstein's view of gravity.

Yet, incredibly, Einstein's belief in the importance of Kaluza and Klein's ideas would ultimately prove correct – though in a way that even he would struggle to comprehend.

With Einstein in his grave and his attempts to unify gravity with electromagnetism in the waste-bin, few physicists were keen to take up the struggle. Instead, attention shifted to unifying electromagnetism with one of the newly found forces: the weak interaction. As Maxwell had shown, the trick lay in finding a mathematical description of both which brought out their hidden similarities. During the 1920s, physicists believed they had found the way to do it, using so-called Quantum Field Theory (QFT).

According to QFT, every fundamental force has its own special messenger particle, and during the 1950s, physicists began to explore

similarities between the carriers of electromagnetism and the weak force: the photon and the W-particle. Digging out the similarities was far from simple – not least because the W-particle is infinitely heavier than the massless photon. Yet by the late 1960s, three theorists – Steven Weinberg and Sheldon Glashow in the US, and Abdus Salam in England – had independently developed theories for these two forces showing that, deep down, they really are just different aspects of a single "electroweak" force. Better still, the unification led to predictions of subtle new effects that could be checked in experiments. When these predictions began to be verified during the 1970s, physicists celebrated the first successful unification since Maxwell's breakthrough over a century earlier.

By then Glashow at Harvard University had embarked on a search for other ways of revealing the underlying unity of forces. In 1973, working with his colleague Howard Georgi, he had found a mathematical format that unified electromagnetic, weak and strong nuclear force. Known as Grand Unified Theory (GUT), it opened up a truly cosmic vision of the fundamental forces of nature, according to which all three of these forces were once part of a single "superforce" that ruled the universe just after the Big Bang. As the universe cooled, the three forces split apart, creating the cosmos we see today.

Again, the theory made predictions, but this time confirmation was harder to come by. Theorists found that the original version of GUT lacked a crucial ingredient – and one that produced another amazing unification. Discovered by theorists in the early 1970s, "supersymmetry" is a mathematical property which unites the particles which make up matter, such as electrons and protons, with those that carry forces, such as photons.

Theorists found that supersymmetry stripped away all the apparent differences of these sub-atomic particles to reveal their basic unity. But they also found that it gave them another, even more profound, clue about the Theory of Everything – one which hinted at how to unify the superforce of GUT to the one remaining outsider: gravity.

Some theorists had already tried bringing gravity into the fold using Quantum Field Theory, which portrayed the force as particles, called gravitons, flitting between masses. Yet, like Einstein, they had all run into horrendous mathematical problems. Even so, the most promising of these heroic failures all included supersymmetry, plus something else: extra dimensions – the very idea that had entranced Einstein half a century earlier. What they lacked was some ingredient that would unify gravity with the other forces without unleashing mathematical demons. It proved to be something so radical even Einstein might have baulked at it: the superstring.[1]

In 1984, theorists John Schwarz at Caltech and Michael Green of London University stunned their colleagues by declaring that they could unify gravity with the other forces without the usual problems – but on one condition: that particles were no longer regarded as mere points, but as tiny entities called superstrings. Far smaller than an atomic nucleus, these thread-like objects also had to possess supersymmetry (hence "superstrings") and exist in ten dimensions.

It was an astonishing claim, and prompted armies of theorists to find out more about superstrings. By the late 1980s it was clear that, despite being a major advance, they weren't the whole story. While there can be only one Theory of Everything, theorists uncovered no fewer than five superstring theories, and no clear way of choosing between them. Superstrings seemed to be just a shadow of something even grander.

In 1995, string theorist Edward Witten of the Institute of Advanced Study, Princeton, unveiled what many now regard as the first view of that ultimate theory, perhaps the Theory of Everything itself.

Witten showed that all five superstring theories are just rough descriptions of a single, overarching idea, which he dubbed M-theory. Some theorists argued that the "M" represented "Mother", "Mysterious" or even "Magic", but its connection with superstrings is clearest if it stands for "Membrane". The five superstring theories then emerge as merely the multi-dimensional "edges" of 11-dimensional membranes, all but 4 of whose dimensions are curled up too small to see.

Today M-theory remains the best candidate yet for what Einstein sought, and far more besides: a single, unified description of not just electromagnetism and gravity, but of the other fundamental forces, and all the particles on which they act.[2]

It is a truly mind-boggling achievement, and it is not over yet. Many of the world's best theorists are digging into the rich mathematical seams of M-theory, looking for answers to the many mysteries that remain. Exactly how and why did all but 4 of M-theory's 11 dimensions curl up too small to see? Can they be detected experimentally? Why is the beautiful unity between forces and particles so hard to uncover?

Even M-theory may ultimately prove to lack the power to answer all these questions. But at the very least, it has given us an awe-inspiring glimpse of the basic unity of the cosmos and everything in it.

Notes

1. While superstring theory and, latterly, M-theory have been the focus of most attention from physicists searching for the Theory of Everything, they are certainly not the only games in town. Indeed, they may not even be the best ones. Some theorists have taken a radically different approach to tackling the key problem of unifying Quantum Theory with Einstein's theory of gravity, General Relativity. Known as loop quantum gravity, it applies quantum theory ideas directly to Einstein's equations, doing away with certain questionable assumptions and approximations made in superstring theory. One consequence is that space and time emerge from the theory in the form of incredibly small "loops", out of which our universe is formed. Many insights from loop quantum gravity have also turned up in superstring theory, which – given the independence and greater rigour of the former – is helping to boost confidence in a quest so far sorely lacking in reality checks.

2. One major problem facing the search for the Theory of Everything is the sheer plethora of solutions to the equations of superstring theory. Some estimates put it at more than 1 with five hundred noughts after it – and each describes an entirely different universe. This is clearly unsatisfactory, as at best only one of these many solutions can be right: the one corresponding to our universe. There is, however, a way around this. Perhaps our universe, with its mix of forces and particles, is just part of a far greater "multiverse" made up of countless universes, each with its own set of laws. If so, then perhaps it simply makes no sense to ask which – if any – of the superstring equations are correct. They might all be accurate descriptions of different parts of the multiverse. For more about the concept of the multiverse, see chapter 24.

Further reading

The Second Creation by Robert Crease and Charles Mann (Quartet, 1997)

Three Roads to Quantum Gravity by Lee Smolin (Weidenfeld & Nicolson, 2000)

The Fabric of the Cosmos by Brian Greene (Penguin, 2004)

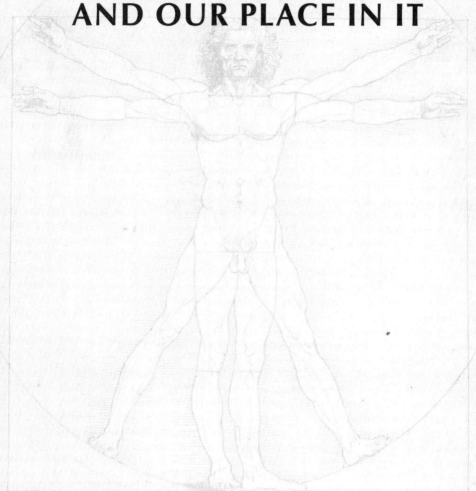

THE UNIVERSE –
AND OUR PLACE IN IT

22
The Big Bang

IN A NUTSHELL

Until around a century ago, most scientists clung to the Aristotelian view of the universe as everlasting and unchanging. In 1917, Einstein found that his new theory of gravity, General Relativity, refused to conform to this ancient view, and was forced to alter his equations. It quickly emerged that he had blundered, and that the universe was actually expanding, as if blown apart in a huge explosion billions of years ago.

Cosmologists have since uncovered compelling evidence for this "Big Bang", and are now trying to discover its cause. Quantum effects known as scalar fields are thought to have propelled an initial period of extremely rapid expansion, known as inflation, and to have triggered the release of radiation and matter we call the Big Bang. The major challenge now is to explain the origin of these scalar fields, and to show that they actually existed at the moment of creation.

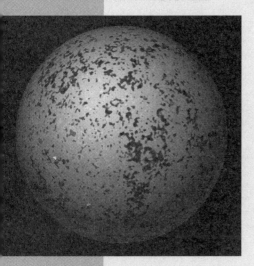

Next time you're looking for something inspiring on TV, find the channel tuning button and press it until you're between channels. Look closely at the blizzard of static that covers the screen. Around 1 per cent of it is microwave energy which started out as heat from the Big Bang almost 14 billion years ago.

Now who says there's never anything on the TV?

Of all the scientific advances of the last century, none is more awe-inspiring – or more perplexing – than the discovery that our universe began with a colossal explosion and has been expanding ever since. It is a conclusion so profound that even

Einstein, the most radical scientist of modern times, struggled to accept it. How could a universe just come into existence from nowhere ? What triggered it – and where did the matter we now see come from?

After decades of effort by both theoretical and observational astronomers, answers are now starting to emerge, based on hard evidence rather than speculation. But they paint a picture of a universe formed by forces whose origins and power are barely less miraculous than the biblical account of the Creation.

It is enough to make one almost nostalgic for the simple view of the cosmos put forward by the Greek

TIMELINE

1917 Einstein applies his theory of General Relativity (GR) to the whole universe, but finds it fails to give the "right" answer of a static, everlasting cosmos.

1922 The Russian theoretical meteorologist Alexander Friedmann finds equations showing GR allows a non-static universe; a doubtful Einstein eventually concurs.

1927 Prompted by red-shift data, the Belgian cosmologist and priest Georges Lemaître argues that GR implies a universe which had a beginning.

1929 American astronomer Edwin Hubble announces discovery of the expansion of the universe, with galaxies racing away from each other as if from an explosion.

1946 George Gamow, Ralph Alpher and Robert Herman begin study of the creation of chemical elements in the Big Bang.

1948 The Steady State Theory is proposed by Herman Bondi, Thomas Gold and Fred Hoyle at Cambridge University, who argue the universe does not change over time.

1950 Fred Hoyle coins the term "Big Bang" in a series of BBC radio lectures; meant to be derisive, it becomes the standard description of the theory.

1964 Engineers Arno Penzias and Robert Wilson at Bell Laboratories detect the cosmic microwave background (CMB), seen as confirming the Big Bang theory.

1970s Soviet theorist Andrei Linde and Alan Guth in the US highlight role of "scalar fields" in triggering inflation in the very early universe.

1992 COBE satellite studies CMB microwave background in unprecedented detail, and begins new era of "high precision cosmology", with reliable estimates of many universal properties.

1995 Studying exploding supernovas, astronomers find convincing evidence for the existence of "dark energy", which is now thought to dominate the universe.

philosopher Aristotle and others over 2000 years ago, according to which the universe is infinite, everlasting and unchanging. Certainly Albert Einstein saw no reason to challenge those assumptions when he published an account of the cosmic implications of his new theory of gravity, General Relativity (GR), in 1917. According to GR, matter curves the fabric of space and time around it, producing effects which we interpret as the "force" of gravity. This was a radical departure from the rather vague view of gravity provided by Newton, according to which the force worked like some sort of invisible elastic. Einstein had shown that GR gave a more accurate account of reality than Newton's concept of gravity, at least within the Solar System, and fully expected it to triumph when applied to the cosmos as a whole.

Yet he quickly ran into a major problem: in their purest, most elegant form, the equations of GR refused to give the "right" answer, pointing instead to a universe that was anything but static. Perplexed by this apparently ludicrous result, Einstein slipped an extra term into his equations, a fudge-factor later called the "cosmological constant". It was the first symptom of the problems Einstein would have grappling with cosmic questions.

The second emerged almost immediately, when the Dutch theorist Willem de Sitter showed that GR allowed an expanding yet unchanging universe. Einstein had hoped GR would have the hallmark of the "ultimate" theory of gravity, and produce a unique model of our universe – namely, the one that actually matches reality. He thus initially thought de Sitter had made a mistake, but failed to find one. Then in 1922, a Russian meteorologist named

Alexander Friedmann added insult to injury by showing that GR allowed a whole host of universes, some of which were neither static nor unchanging. Again Einstein suspected a blunder, but couldn't find one.

The biggest blow came in 1927, when a Belgian cosmologist and priest Georges Lemaitre reached the same theoretical results as Friedmann – and pointed to hard evidence that the universe really was expanding. Astronomer Vesto Slipher of the Lowell Observatory in Arizona had found that the light of blob-like "nebulae" in the night sky often showed a curious shift towards the redder – longer-wavelength – end of the spectrum. Slipher thought he was witnessing the so-called Doppler Effect, familiar from the increasing wavelength of sound of police sirens as they speed away. In fact, he was seeing something far more profound: the stretching of light by the expansion of the whole universe.

The first hints of this came in 1924, when Edwin Hubble of the Mt Wilson Observatory in California showed that nebulae were actually vast galaxies far beyond our own.[1] Five years later Hubble published one of the most important results in modern science: a graph showing that the "red-shifts" of these distant galaxies increased steadily with distance. Now called Hubble's Law, this was precisely the kind of behaviour predicted by GR for an expanding universe.

Friedmann and Lemaitre had been right all along, and Einstein was left knowing that by forcing his equations to conform to ancient beliefs, he had missed out on making the most astounding prediction in scientific history.

It fell to Lemaitre to point out the no less astounding implication: that the universe must once have burst into existence in some titanic explosion. He went on to be hailed as the father of the Big Bang theory. Einstein, meanwhile, came to regard his missed opportunity as the greatest blunder of his life.

JARGON BUSTER

Red-shift effect: A consequence of the expansion of the universe, in which radiation from distant sources is "stretched" towards longer wavelengths. In the case of light from galaxies, this appears as a shift towards the red end of the spectrum; hence the term "red shift".

Cosmic Microwave Background (CMB): Almost 14 billion years after the Big Bang, the universe still contains a faint remnant of the colossal heat and radiation created during its birth. It is detectable today as a more or less constant "hiss" of microwaves across the whole night sky – this being remnant of infrared radiation – heat – whose wavelength has been stretched into microwaves by the cosmic expansion.

Hubble constant: Often also called the Hubble Parameter (H_o) because it is not actually fixed at all, the Hubble constant measures the rate at which the universe is currently expanding. Current measurements imply that the universe began expanding around 13.7 billion years ago.

Scalar field: A kind of quantum force field likely to exist in the very early universe. Predicted to have anti-gravitational effects, a scalar field allows the rapid expansion of the universe known as inflation, and its collapse triggers the release of radiation and matter characteristic of the Big Bang.

Was the Steady State Theory right after all?

The demise of the Steady State Theory in the late 1960s saddened many cosmologists, who regarded it as a beautiful theory killed by the ugly fact of the Cosmic Microwave Background (CMB). One of its originators, Sir Fred Hoyle, continued to defend it – or, at least, a modified version of it – up until his death in 2001. In *A Different Approach to Cosmology*, published the year before, Hoyle and leading astrophysicists Geoffrey Burbidge and Jayant Narlikar argued that all the criticisms levelled at the Steady State Theory in the 1950s and 1960s were either misplaced, or had plausible explanations.

Few showed much interest, however, and the Big Bang still rules the roost. Yet it is not without its own problems – and, ironically, these may yet resurrect the most beautiful aspects of the Steady State. The recent discovery that the universe is filled with "dark energy" has revealed that the effect of ordinary matter on the cosmic expansion is much less than originally believed. Instead, trillions of years from now the universe will be propelled solely by dark energy – and its behaviour will then be identical to that predicted on the basis of the Steady State model.

It soon became clear that the cosmos could play games with astronomers as well as theoreticians. Early measurements of the cosmic expansion rate implied that the universe was born only a few billion years ago – thus making it younger than the Earth. In 1948, three Cambridge theorists – Hermann Bondi, Thomas Gold and Fred Hoyle – found an ingenious way out of this paradox. They showed that Hubble's Law was consistent with a universe that was in a "steady state", its expansion rate and density never changing. Mathematically, that implied that despite appearances the universe had actually been born infinitely long ago.

The Steady State model thus dodged the cosmic age problem, and impressed many with its elegance. Yet to keep the density of matter constant, it had to allow matter to leak into the universe from nowhere. Its originators pointed out that this was hardly less implausible than the entire universe suddenly appearing from nowhere in what Hoyle derisively called a "Big Bang".

By the late 1950s, new measurements showed that the age of the universe was comfortably greater than that of the Earth, undermining the *raison d'être* of the Steady State model. But the real death-blow came in 1964, with the most important cosmic discovery since Hubble's Law. Two American engineers were trying to find the source of a curious microwave "hiss" detected by a horn-shaped radio aerial at Bell Labs in New Jersey. It seemed to come from beyond the Earth, but its real significance only emerged after the engineers talked with astrophysicists at Princeton University – who told them the hiss had precisely the form expected of red-shifted heat radiation left over from the Big Bang.

The discovery of this Cosmic Microwave Background (CMB) is now regarded as the best evidence that the universe began in a colossal fireball.[2] But it is not the sole evidence. Over 50 years ago, the Russian-American astrophysicist George Gamow and his colleagues realised that the heat of the Big Bang would trigger nuclear reactions that could create chemical elements.[3] They calculated that the universe should contain around three parts hydrogen to one of helium – which astronomers later confirmed.

During the 1970s, theorists began to grapple with quantum effects likely to exist in the small, intensely hot, early universe. Attention focused on so-called scalar fields, which cosmologists in the Soviet Union and the US showed could produce an incredibly rapid burst of cosmic expansion called inflation. Better still, these anti-gravity fields could be made to dump their energy into the new-born universe, providing just the kind of heat and matter needed for a Big Bang.

That at least was the theory, and support for it started to emerge in 1992, with the advent of a new era in cosmology. The orbiting Cosmic Background Explorer (COBE) satellite examined the microwave radiation filling the universe with unprecedented precision, and gave cosmologists insights once only dreamed of. COBE showed that the original fireball has now cooled to a temperature just 2.73 degrees above Absolute Zero. It also showed that the CMB had technical characteristics broadly in line with inflation theory.

Since then, ground-based observations and results from NASA's orbiting Wilkinson Microwave Anisotropy Probe (WMAP) have provided more detail. Further insights into inflation are expected from the European Space Agency's Planck satellite, due for launch in 2007. Taken together, the data so far paint a picture of a baby universe that burst literally from nowhere around 13,700 million years ago. Filled with scalar fields, it then suddenly inflated from something around a billion times smaller than a proton to the size of a grapefruit in around 10^{-32} seconds. Then, just as suddenly, the scalar field collapsed, releasing energy as radiation and matter in what we now call the Big Bang.

Thus it now seems that the Big Bang was not the start of everything after all, but was merely the result of effects already at work in a new-born universe. Understanding those effects

Will there be another Big Bang?

Current data suggest that the universe was created in a Big Bang around 13.7 billion years ago, and will expand forever – propelled by the invisible power of "dark energy". This suggests our universe is doomed to just fade away. Yet according to cosmologists Sean Carroll and Jennifer Chen of the University of Chicago, the reality may be very different. They point out there is a chance the dark energy will undergo some fearsome random fluctuation, triggering another Big Bang. It's an extremely remote possibility – just 1 in 10 to the power 10 to power 56 – but it's not zero either, and may explain the last Big Bang.

will require nothing less than a Theory of Everything, capable of explaining all the forces of nature and the particles on which they act – and even the origin of space and time.

It is a daunting prospect, but theorists hope to tap into a potent new source of insight: gravitational waves. Predicted to exist by Einstein in 1916, these ripples in the very fabric of space and time are thought to be triggered by the most violent cosmic events, such as supernova explosions and the collision of black holes. Astronomers are now planning huge space-based observatories capable of detecting the gravitational waves created during the Big Bang, inflation – and perhaps even the actual moment of creation itself.[4]

So far, no one has detected gravitational waves; their existence remains the one major prediction by Einstein still awaiting confirmation. If, as most scientists believe, Einstein was right and gravitational waves really do exist, they may well be the key that unlocks the ultimate cosmic mystery he tried so hard to solve.

Notes

1. Despite centuries of observations by astronomers, the true nature of the blobs of light called "nebulae" dotted around the night sky remained unclear until the 1920s. Then the American astronomer Edwin Hubble used observations of a special type of variable star to work out the distance to one such nebula: M.31 in the constellation Andromeda. His result of 900,000 light-years showed that M.31 was actually a vast galaxy of stars lying far beyond our own. The first galaxies are thought to have been formed from the primordial "lumpiness" in the matter created during the Big Bang. The latest observations of the Cosmic Microwave Background suggest the first galaxies formed just 200 million years after the birth of the universe.

2. A form of electromagnetic radiation with wavelength between 1 mm and 30 cm, microwaves lie between infrared radiation and radio. This has led to applications ranging from cooking food to mobile phones and satellite communication – with an aerial designed for the latter revealing the reality of the Big Bang.

3. Born in 1904, Russian-born Gamow was among the most original scientists of the last century. As a graduate student he showed how the newly discovered uncertainty principle could explain radioactivity, and he went on to make key contributions to cosmology, astrophysics, and even biochemistry, working on the genetic code of DNA.

4. To detect gravitational waves, scientists in the US have constructed the £200 million Laser Interferometer Gravitational-wave Observatory (LIGO). There are in fact two sites, in Louisiana and Washington, to minimise false alarms, each using lasers shone down L-shaped tubes 5km long to detect tiny movements due to gravitational waves.

Further reading

Cosmology: A Very Short Introduction by Peter Coles (Oxford, 2001)

Cosmological Physics by John A. Peacock (Cambridge, 1999)

The Universe in a Nutshell by Stephen Hawking (Bantam, 2001)

Fabric of the Cosmos by Brian Greene (Penguin, 2004)

23
Dark Energy

IN A NUTSHELL

Ever since the time of Isaac Newton, it has been an article of faith among scientists that the most important force in the cosmos is gravity. Until recently, astronomers were convinced that ever since the Big Bang, the universe has been expanding at a decreasing rate under the influence of gravity. The big question was whether gravity was strong enough to eventually halt the expansion, and bring about a cataclysmic "Big Crunch". In the early 1990s, astronomers set about answering this question by studying the light from distant supernovae – exploding stars on the edge of the visible universe. By 1998, it became clear that, far from slowing down, the cosmic expansion is actually accelerating. This finding, since confirmed by other methods, has revealed the existence of "dark energy" in the universe, which propels the expansion rate against the pull of gravity.

So far, attempts to develop a theory of dark energy have failed, though it is thought to be linked to so-called quantum vacuum fluctuations – sub-atomic effects whose existence has been demonstrated in the laboratory. The ultimate explanation is expected to emerge from the long-sought Theory of Everything, which will reveal the hidden connections between the sub-atomic and cosmic worlds.

It was a telephone call that astronomer Robert Kirshner of Harvard University won't forget in a hurry. A colleague had called him to discuss new observations of exploding stars on the very edge of the visible universe. And Kirshner's colleague was telling him something he didn't want to hear: that the universe is in the grip of a force that no one can explain.

Usually scientists would be delighted to make such an astonishing discovery. But this was different. This was a force that had once made a fool of Einstein himself. What concerned Kirshner and his colleagues was the distinct possibility that it was about to make fools of them as well.

That was in December 1997. Since then, it's become clear that those

TIMELINE

1917   Einstein uses his new theory of gravity, General Relativity, to study the mystery of the cosmos, and introduces the so-called cosmological constant.

1927   German physicist Werner Heisenberg publishes the Uncertainty Principle, showing that even empty space is seething with "dark energy".

1929   American astronomer Edwin Hubble discovers that the universe is expanding, showing Einstein's cosmological constant was unnecessary.

1948   Dutch physicist Hendrick Casimir predicts that empty space can generate an attractive force between two metal plates, pushing them together.

1981   Alan Guth of the Massachusetts Institute of Technology suggests that the early universe was "inflated" by a kind of primordial dark energy.

1984   Stephen Hawking puts forward a theory which he claims can explain why there is no dark energy in today's universe.

1988   Nobel Prize winning physicist Steven Weinberg describes lack of understanding of the cosmological constant and dark energy a "veritable crisis" for physics.

1996   Steve Lamoreaux of the University of Washington, Seattle, succeeds in measuring and confirming the existence of the incredibly weak Casimir Force.

1998   Teams of astronomers studying light from distant supernovas unveil first hard evidence for dark energy propelling the expansion of the universe.

2001   NASA launches the Microwave Anisotropy Probe (MAP), which two years later produces results showing that dark energy dominates the cosmos.

2003   John Tonry and colleagues at the University of Hawaii show that dark energy wrested control of the universe from gravity around 7 billion years ago.

observations of exploding stars weren't playing tricks. Other astronomers have made similar observations, and they have been backed up by data gathered from space. There is now little room for doubt. The universe is not just expanding: it is expanding at an ever-faster rate, propelled by a bizarre force that comes literally out of nowhere.

Its name captures its mysterious nature perfectly: "dark energy".

Until now it has been an article of faith among astronomers that the most important force in the cosmos is the same one that Isaac Newton saw pull an apple from a tree almost 350 years ago. But no longer. Astronomers now believe that around 7 billion years ago, gravity lost control of the fate of the universe to this mysterious dark energy. The problem is that no one can say why, or how. Apart from its presence, virtually nothing is known about dark energy – the biggest enigma in all science.

Small wonder, then, that scientists have mixed emotions about the discovery of dark energy. Their reluctance becomes all the more under-standable, given the fate of the first person to tangle with it: Albert Einstein. Already famed for his revolu-tionary theory of Special Relativity and insights into Quantum Theory, in 1915 Einstein unveiled his masterwork: a radically new theory of gravity, the first since the time of Isaac Newton. Known as General Relativity (GR), it envisaged gravity not as a mysterious influence that somehow extended over the vacuum of space, but as the warping of the very fabric of space and time by everything in it, from galaxies to apples.

Like Newton before him, Einstein had shown that his theory could explain otherwise baffling facts about the movement of the planets, and predicted effects that were later triumphantly confirmed. But Einstein also shared Newton's keenness to push his theory further, using it to understand the whole of creation. In 1917, he applied the equations of GR to the cosmos as a whole, to see what fresh light they could cast on the nature of the universe. And his equations had a big surprise for him: they pointed to a universe that was expanding.

At the time scientists believed that the universe was static and unchanging, a view backed by astronomical observations. The very idea of an expanding universe seemed preposterous: expanding from what, and into what? Confronted by equations that failed to agree with this "common sense" view of the cosmos, Einstein felt he had no choice but to introduce a fudge-factor to force his theory into producing a static, unchanging universe.

Often called the cosmological constant or simply "lambda", this fudge-factor amounted to a new force at work in the universe, cancelling out the effect of gravity. Einstein came bitterly to regret his decision. By the late 1920s, astronomers had found that they had been wrong: the universe really is expanding – making Einstein's fudge-factor unnecessary. If only he had kept faith with his original equations, Einstein could thus have made the greatest prediction in the history of science: that the universe was created in an explosive "Big Bang", and has been expanding ever since.

Years later, Einstein allegedly told a friend that fiddling with the equations of GR was the "biggest blunder of my life". Yet by then discoveries had been made that would ultimately show that the truth was stranger than Einstein could have imagined.

The first hints emerged in 1927, when the German physicist Werner Heisenberg unveiled the famous Uncertainty Principle. This showed that it is impossible to have perfect knowledge of the sub-atomic world. Not surprisingly, the link with cosmology wasn't immediately

JARGON BUSTER

Uncertainty Principle: A relationship between certain properties of sub-atomic particles – such as their position and momentum – which implies that the more certain one property is known, the less certain the other becomes. The uncertainty in the energy content of empty space is thought to be responsible for "dark energy".

Quantum vacuum fluctuations: A vacuum is supposed to be completely devoid of content, but in reality the Uncertainty Principle implies it's seething with energy, particles and anti-particles constantly flitting in and out of existence. The effects of these quantum vacuum fluctuations have been measured in the laboratory, and may be linked to "dark energy".

Inflation: The explosive expansion of the universe moments after its creation, propelled by the huge amount of dark energy then thought to exist. While inflation solves some otherwise baffling cosmic mysteries, explaining why inflation took the form it did is itself a major mystery.

Casimir Force: A force which appears between two metal plates brought very close together, created by quantum vacuum energy being greater outside the plates than within, and "hammering" on the plates, pushing them together. First predicted in 1948, it is too feeble to measure except when objects are brought within a fraction of a hair's breadth of each other.

The vacuum effects in your office

Despite their apparently esoteric nature, quantum vacuum effects crop up in many everyday circumstances. Electronic circuits are prey to the random "noise" generated by random vacuum energy fluctuations, which put fundamental limits on the level to which signals can be amplified. The same effect is used by ERNIE – the Electronic Random Number Indicator Equipment – to generate the random numbers used to pick winning Premium Bonds each month. Even fluorescent strip lighting owes its operation to the causeless, random energy fluctuations of the vacuum state. The mercury vapour atoms are excited by the electrical discharge in the tube, their spontaneous emission of photons is triggered by vacuum fluctuations knocking them out of their unstable energy state. In other words, every time you switch on the office lights, you are triggering effects related to those now propelling the expansion of the universe.

apparent. It is there, nonetheless – in the fact that, according to the Uncertainty Principle, it's impossible to know the precise amount of energy in space at any given time. Even in an apparently utterly empty vacuum, there will always be so-called quantum vacuum fluctuations of energy taking place within it. And, crucially, the effects of these fluctuations aren't confined to the sub-atomic world – as the Dutch physicist Hendrick Casimir first showed in 1948. He calculated that if two metal plates are brought very close to one another, they exclude some forms of quantum vacuum fluctuation. With fewer fluctuations trapped between the plates than outside, the plates are pushed together, as if by an invisible force-field. So feeble is this force that its existence was only properly confirmed in 1996. Even so, Casimir had served warning on scientists: empty space is capable of springing big surprises.

Just how big did not become clear until the late 1970s. By then, cosmologists were confident that the universe had begun in a Big Bang; now they wanted to figure out its causes. Working back from the current expansion, it seemed clear that the universe must once have been very small – so small, in fact, that it must have been governed by Quantum Theory, the bizarre rules of the sub-atomic world.

In 1978, a young American theorist named Alan Guth sat in on a lecture on cosmology, and began to see the links between the cosmos and his own area of expertise: particle physics. Guth had been working on the forces at work in the sub-atomic world: the so-called strong and weak nuclear forces, and electromagnetism. Following a plan first hatched by Einstein, theorists like Guth suspected that all these forces are just different facets of a single "unified field". Calculations suggested that the three sub-atomic forces would indeed merge together – but only at incredibly high temperatures. Guth, among others, realised that just such temperatures would have occurred moments after the Big Bang.

In 1981, he published a revolutionary paper spelling out the implications of these "grand unified theories" (GUTs). According to Guth, they suggested that an extremely powerful cosmic force could have emerged during the Big Bang, triggering an incredibly rapid expansion of the universe. While the details of Guth's idea turned out to be flawed, other theorists seized on the basic idea of "cosmic inflation", not least because it solved some awkward technical problems about the Big Bang.[1]

But there was a sting in the tail. The effect of Guth's new force could be captured as a simple addition to Einstein's theory of gravity – and it was identical to Einstein's bane: the cosmological constant. Astronomers now had to explain why the once huge cosmological constant had vanished completely just moments after the Big Bang. The problem was made worse by the fact that rough estimates of the current size of the cosmological constant gave simply ludicrous results – typically 120 powers of 10 larger than observations allowed. During the 1980s, some of the world's most celebrated theorists, including Stephen Hawking at Cambridge University, tried to come up with arguments to explain why the cosmological constant had completely vanished. None of them were very convincing, prompting the American Nobel Prize winning physicist Steven Weinberg to declare that the cosmological constant represented a "veritable crisis" for physics.

Ten years later, in 1998, that crisis took a dramatic new turn. Astronomers began to unveil evidence that everyone had been wrong about the cosmological constant. Far from having vanished just after the Big Bang, it appeared to be still at work in the universe. Indeed, it seemed to have wrested control from gravity, propelling the cosmic expansion at an ever greater pace.

These stunning revelations emerged from studies of the death-throes of giant stars at the very edge of the visible universe. Known as supernovae, they can briefly outshine the combined light of an entire galaxy – making them ideal probes of deep space. In the mid-1990s, two teams of astronomers began using these distant beacons to study the rate of expansion of the universe. The idea was simple enough: to find out whether the universe is expanding so fast that it will expand forever, or if it would one

How will the universe end?

The discovery of dark energy has profound implications for the future of our universe. Cosmologists once thought that the fate of the cosmos would be determined purely by the density of matter in the universe. If this were high enough, gravity would eventually halt the expansion, prompting a collapse back to a cataclysmic Big Crunch; otherwise it would expand forever. It is now clear, however, that dark energy has been the dominant force in the cosmos for the last seven billion years, and is driving the acceleration at an ever-increasing rate. As such, the universe looks set to expand forever. If so, our universe also looks set to become a very desolate place. For example, our galaxy will find itself entirely isolated within the next 100 billion years, the rest of the cosmos having been transported to regions so far away their light can never reach us.

day slow down under the force of its own gravity, come to a halt – and collapse back towards a cataclysmic Big Crunch.

Either way, two international teams of astronomers set out expecting to find clues in the fading light from the supernovae. What they actually found left them reeling – and led to that phone-call between Robert Kirshner at Harvard and his colleague Adam Riess. As members of the High-z Supernova Search team, Kirshner and Riess had been gathering telescopic observations, looking for signs of a change in the cosmic expansion rate. By December 1997, they were beginning to find it. But far from showing a universe slowing down under its own gravity, the supernovae pointed to a universe that was speeding up, propelled by an invisible form of energy – soon dubbed "dark energy".

Similar results were announced by the rival Supernova Cosmology Project, led by Saul Perlmutter at the Lawrence Berkeley National Laboratory in California. And in the last few months, the supernova results have been confirmed by the most detailed study ever made of the heat left over from the Big Bang. Compiled by NASA's orbiting Wilkinson Microwave Anisotropy Probe (WMAP), the data tell the same story as the supernovae, and the implications are disturbing indeed. Einstein, Hawking and the other brilliant theorists were wrong: the most potent force in the universe is not gravity, but "dark energy" – in the form of the cosmological constant.

With its reality beyond doubt, theorists are now trying to figure out the consequences of dark energy. They could be dramatic: in 2003, one team of theorists suggested that the ever-accelerating cosmic expansion may eventually tear apart space and time in a so-called "Big Rip".

Others are struggling simply to explain the existence of dark energy.[2] It seems certain that its source is the quantum vacuum fluctuations predicted by Heisenberg's Uncertainty Principle. Calculations show these can produce dark energy with the necessary "anti-gravitational" effect. The big problem lies in explaining the precise level of dark energy now observed – currently thought to account for around 75 per cent of the total energy in the universe.

Most theorists think the ultimate answer will emerge from the long-sought Theory of Everything, which will account for all the forces of Nature in a single set of equations. In the meantime, the existence of dark energy remains a truly cosmic enigma – and one that may only be resolved by the twenty-first-century equivalent of Albert Einstein.

Notes

1. One such problem lay in the observed density of matter in the universe. During the 1960s and 1970s, astronomers had used a variety of techniques to discover that the density of the universe was fairly close to the crucial dividing line between a universe so dense that it would eventually recollapse into a Big Crunch, and one tenuous enough that it would continue to expand forever. But during the 1970s, the American physicist Bob Dicke at Princeton University pointed out that the fact that the density was even remotely close to this dividing line was itself remarkable. Working backwards in time, he showed that it meant that the universe must have been within one part in one million billion of this critical value just one second after the Big Bang in order to end up so close to the dividing line today. Many physicists could not believe that this could have been an accident. One of the triumphs of inflation theory was to show how this balancing act was achieved: put simply, it is the result of the colossal amount of expansion that took place just after the creation of the cosmos.

2. There are currently several candidates for the source of dark energy. In the late 1960s, when astronomers thought (wrongly) that their observations had found hints of a cosmological constant, the Soviet astrophysicist Yakov Zel'dovich suggested that dark energy might be due to gravitational interactions between vacuum particles. He even came up with a formula for the size of the effect, but it required a certain value for a sub-atomic mass to be entered, and no-one has come up with a good explanation for the size of that mass.

Another candidate is so-called "quintessence" (named after the fifth element of the Ancient Greek cosmos, after Earth, Air, Fire and Water). The strength of this form of dark energy can vary with space and time, allowing it to operate both at the time of inflation during the Big Bang, and also in today's universe. Theorists hope that more detailed observations of the universe from successors to WMAP and from orbiting telescopes such as the planned US Supernova/Acceleration Probe (SNAP) will allow them to identify the true source of dark energy.

Further reading

The Origin of the Universe by John D. Barrow (Weidenfeld & Nicolson, 1994)

The Last Three Minutes by Paul Davies (Weidenfeld & Nicolson, 1994)

Just Six Numbers by Martin Rees (Weidenfeld & Nicolson, 1999)

24
Parallel Universes

IN A NUTSHELL

Philosophers have argued for thousands of years that the universe must – by definition – be infinite in extent. Recent astronomical observations strongly support that view, implying that what we call our universe is merely part of something literally infinitely larger – a "multiverse" – where every possible permutation of events and conditions exist. An infinite number of these parallel universes are utterly unlike our universe, where different laws of physics hold sway. However, an infinite number of them have conditions just right for the emergence of life – one of which is the universe in which we live. While all these "parallel universes" lie beyond the reach of conventional means of communication, many scientists think their presence reveals itself in subtle effects, such as interference patterns created even by individual photons of light or sub-atomic particles. Some even insist the existence of parallel universes can be put to practical use via so-called quantum computers, now being built in laboratories around the world. These are predicted to be far more powerful than any computer now in existence, as they allow calculations to be carried out simultaneously in a vast number of parallel universes.

As he set up the experiment at his parents' house, Geoffrey Taylor may well have wondered why he was even bothering, for surely the outcome was obvious. Still, his Cambridge University mentor Sir Joseph Thomson had asked him to do it anyway, so he went ahead, setting the needle in its holder, switching on the light-source, and adjusting its filter to control the amount of light getting through. He then exposed the photographic plate, checked everything was working – and went on holiday.

When Taylor returned, he made a stunning discovery. Examining the photographic plate, he could see faint bands of light and dark – the unmistakable pattern of light rays interfering with each other. The same pattern had been seen by many other physicists using similar apparatus, but its

TIMELINE

56 BC	Roman poet-philosopher Lucretius completes *De Rerum Natura* (The Nature of Things), arguing for an infinite universe.
1909	Cambridge physicist Geoffrey Ingram Taylor shows that single photons can create interference effects – hinting at parallel universes.
1915	Einstein publishes his theory of gravity, General Relativity, linking matter to the shape and extent of the universe.
1925	Erwin Schrodinger publishes his quantum wave equation, with its weird implication that particles exist in myriad different states simultaneously.
1957	Princeton student Hugh Everett III puts forward "Many Worlds Interpretation" of quantum theory, in which reality consists of an infinitude of possible universes.
1960	Andy Nimmo of the British interplanetary society coins the term "multi-verse" while giving talk on Everett's ideas.
1985	Oxford physicist David Deutsch proves the existence of universal quantum computers, able to perform any possible calculation.
1997	Astronomer Royal Sir Martin Rees suggests the multiverse may explain why our universe is so well suited for life.
1997	Deutsch argues in *The Fabric of Reality* that interference effects demonstrate the existence of trillions of parallel universes.
1998	Team led by Isaac Chuang IBM, San Jose, carry out first ever quantum computation, using chloroform molecules.
2003	Studies of heat left over from the Big Bang by NASA's WMAP probe provide best evidence yet that the universe really is infinite.

appearance on Taylor's photographic plate was truly amazing. For in his experiment he had exposed the needle to an incredibly low level of light – equivalent to a candle placed about a mile away. At such feeble levels of illumination, only individual particles of light, or photons, were striking the needle one at a time. The very idea of "interference" thus no longer made sense: there simply weren't enough photons around to do it.

Thomson and Taylor had fully expected the photographic plate to show no signs of interference effects, even after weeks of exposure. Yet as they studied the plate that day in 1909, even Thomson – who had just won a Nobel Prize for discovering the electron – had to admit he was baffled.

Physicists have debated the implications of "single-photon interference" ever since. But now some believe it constitutes hard evidence for something once thought the preserve of science fiction novels: the existence of parallel universes.

The basis of this astounding claim lies in a controversy that began over 2000 years ago, and connects Taylor's simple experiment with effects of truly cosmic magnitude. Philosophers have long recognised that the very definition of a "universe" leads to some remarkable conclusions. As long ago as 56 BC, the Roman philosopher-poet Lucretius argued that if the universe were only finite, it would mean that at some point there must be a boundary beyond which lay – well, what? As the universe is, by definition, everything, it must include everything beyond the boundary as well, in which case it cannot be finite after all.[1]

The idea that the universe must be infinite went more or less unchallenged until around a century ago, when Einstein put forward his theory of gravity, known as General Relativity (GR). This revealed a stunning new possibility: that the universe could be finite, and yet not possess a boundary. As a rough analogy, imagine the three-dimensional space of a finite universe curled up to form the surface of a ball.

Now imagine ants crawling over the ball: no matter how long they explore its surface, they never encounter a boundary – despite the fact its surface area is finite.

According to Einstein's theory, the universe can be "finite yet unbounded" if it contains sufficient energy and matter. The latest astronomical observations strongly suggest, however, that Lucretius was right after all: the evidence from distant stars and the heat left over from the Big Bang both point to a universe that really is infinite in extent. Even so, we can only see a tiny part of our unimaginably vast cosmos – the part from which light has been able to reach us in the 13 billion years or so since the Big Bang.

Despite this apparent barrier to knowledge, it is possible to say with absolute certainty what is going on out there in the infinite universe: in a word, everything. Somewhere out there, countless trillions of light years away, another version of you is reading this article while sitting on a train. Somewhere else there is another you, who is a Hollywood movie star, or a Nobel Prize winner – or both.

An infinite universe, in short, contains an infinite number of every conceivable possibility (and inconceivable ones as well). In other words, it contains an infinite number of parallel versions of the universe we can see.

This bizarre yet ineluctable conclusion has led scientists to coin a new word for the truly infinite universe, with its plethora of possibilities: the "multiverse". The term "universe" is now used to describe just the tiny part of the multiverse we can actually observe. The distinction helps resolve some otherwise puzzling questions. Why, for instance, is our universe and its laws just right for the existence of life? Some argue it is because it was specially made for us by a benevolent creator. Yet on the multiverse view, we simply inhabit one of an infinite number of parallel universes suitable for life.

JARGON BUSTER

Single particle interference: A bizarre effect first observed almost a century ago in which even single photons of light or sub-atomic particles exhibit interference effects. Physicists believe the effect reveals the existence of countless parallel universes, whose particles interfere with those in our own universe.

Quantum computing: A form of computing that exploits the fact that sub-atomic particles can exist in combinations – "superpositions" – of different states at the same time. By expressing data in terms of these multiple states, quantum computing allows vastly complex problems to be dealt with in parallel, and thus at enormous speed.

Multiverse: The true "universe", of which ours is just a tiny region. Infinite in extent, the multiverse contains every conceivable form of universe, including an infinite number just like ours. Most are, however, utterly different, containing different laws of physics – virtually all of which are incompatible with the existence of life.

Many Worlds Theory: A view of Quantum Theory according to which sub-atomic particles retain all their possible states on being observed, rather than discarding all but the one actually seen. These other states exist in universes parallel to our own, their existence revealed through the famous "uncertainty" that accompanies attempts to measure the precise properties of sub-atomic systems.

Creating parallel universes

Current theories of the origin of the universe have cast light on possible ways of creating parallel universes. The Big Bang is now thought to have been characterised by a very brief period of extremely rapid expansion known as inflation, propelled by a so-called "scalar field" (see chapter 22). There are many different kinds of scalar field, all with different properties, which raises the question of why our universe ended up being inflated in the way it did. In the early 1980s the Russian cosmologist Andrei Linde proposed that our universe may in fact represent just one part of an infinitely large multiverse in which all the variants of scalar fields are manifest. This "chaotic inflation" model thus suggests that we are surrounded by a plethora of other universes, all with their own peculiarities of content and dynamics.

Many physicists now think the existence of parallel universes can do much more, however. They believe it resolves some of the most baffling puzzles in all science, thrown up by experiments performed a century ago into the nature of mass and energy. These revealed that light, heat and other wave-like forms of radiation also had particle-like properties, their energy coming in bundles known as "quanta". On the other hand, particle-like entities such as electrons were found to have wave-like properties.

More bizarre still were the results of experiments like that conducted by Taylor in his parents' house in 1909, where light continued to show effects like interference, even when only single photons were involved. Many of the most brilliant physicists of the last century set about finding ways of accounting for these strange effects, and the result became known as Quantum Theory.

In 1925, the Austrian physicist Erwin Schrodinger put forward an equation that seemed to capture the essentials of these new findings. At first glance, it looked like a fairly standard equation of the sort used by physicists to describe wave-like behaviour. Lurking within it, however, was something very strange indeed. Put simply, the Schrodinger Equation implied that every particle could be described by a vast number of waves, each reflecting a different possible state for the particle. Yet this was surely absurd: in the real world each particle comes in just one state.

The problem of how to get rid of all the surplus states perplexed Schrodinger and other top physicists for decades. Eventually most of them adopted an idea put forward by the Danish physicist Niels Bohr, known as the Copenhagen Interpretation. According to this, the act of observing a particle rids it of all but one of the myriad possibilities – though quite how or why was unclear.

In 1957, a graduate student at Princeton named Hugh Everett III defied this consensus with a daring proposal: that the vast number of different waves bore witness to the simultaneous existence of the particle in a vast number of parallel universes alongside our own. According to Everett's Many Worlds Interpretation, the reason we see only one of the myriad possible states is simply because there is just one state per parallel universe – and we only observe the one allotted to the universe in which we exist. The presence of the other universes does have an effect on what we see, however: the waves within them affect those in our own universe, creating interference effects – even in the case of single particles. According to the Many Worlds Interpretation, then, Geoffrey Taylor was right to be astonished by what he had observed in his experiment: he had been the first to detect the existence of parallel universes.

While many theorists now think Quantum Theory is best understood in such terms, most workaday scientists are happy just to use Quantum Theory, without worrying about such meta-physical ideas.[2] Yet some physicists believe this view of Quantum Theory has real, practical value – and that the existence of parallel universes may have commercial pay-offs.

In 1985 David Deutsch of Oxford University published a ground-breaking paper suggesting it might be possible to build a computer which could solve any conceivable problem with incredible speed, by exploiting the existence of parallel universes.

In conventional computers, problems are first converted into "bits" – ones and zeros which can be handled by microprocessors as "on" and "off". Calculations are then a matter of storing and shuffling these bits around as fast as possible. Deutsch's paper opened up the prospect of a quantum computer, which takes advantage of the multiple states in which particles can simultaneously exist. In quantum computing, problems would first be converted into a new type of bit – a "qubit" – which is neither wholly one or zero, but a mixture of the two. This allows a single qubit to act like two different bits simultaneously – poten-tially doubling the speed of calculation. In effect, said Deutsch, a quantum computer performs calculations in parallel universes at the same time. The increase in speed could be dramatic. For example, a quantum computer operating on just 100 qubits is equiv-alent to a conventional computer using 2 to the power 100 ordinary bits – that is, a million million billion billion bits, far greater than the capacity of all the world's supercomputers combined.

Bridging the gap between parallel universes

The concept of parallel universes is now taking centre stage in the physics both of the largest scale – cosmology – and the sub-atomic scale, via Quantum Theory. Such convergence of thinking is obviously suggestive, but unifying the two visions of parallel universes poses problems. For example, cosmologists view the different "universes" that make up the infinitely vast multiverse as isolated from us by their distance and dynamics. On the other hand, the parallel universes in the "Many Worlds" Interpretation of Quantum Theory appear to be much "closer" to us in some sense, their presence revealing itself through quantum effects such as interference and uncertainty. Perhaps the long-sought-after Theory of Everything will reveal the underlying unity of these two visions of parallel universes.

Deutsch's paper sparked an explosion of interest in quantum computing. Computer scientists set about designing suitable software, while physicists searched for ways of creating qubits. Attention focused on using sub-atomic particles like protons, which can exist in two different states simultaneously.

Researchers soon discovered a major problem, however: qubits are exquis-itely sensitive, and the information they encode is easily garbled. In 1998, a team led by Isaac Chuang of IBM in San Jose showed that one way around the problem was to use a vast number of qubits, in the form of trillions of protons in a small tube of water. The idea was that at least a few qubits would survive the computation intact. And it appeared to work. Chuang and his colleagues converted a short list of data into qubits, and used a special quantum computing program to sort the list into ascending order, "reading" the result by applying magnetic fields to the tube of water. In 2001, the team used the same technique to factorise 15 into its prime factors of 3 and 5.

While both problems were almost risibly trivial, they did show that quantum computing was not just a theoretical possibility. What remains unclear is whether quantum computing can ever achieve its full, awe-inspiring potential of solving enormous problems at lightning speed.

Scientists working on quantum computing believe commercial machines are a decade away at least. If they succeed, it will surely be one of the most mind-boggling technological advances of all time, exploiting the existence of the parallel universes that exist all around us.

Notes

1. The attempts of philosophers to uncover the nature of the universe have long been derided by physicists. "Is not all philosophy like writing in honey?" asked Einstein. "It looks wonderful at first sight, but when one looks again, it is all gone. Only mush remains." Yet the fact remains that surprisingly simple metaphysical arguments can lead to insights about space and time that physicists have reached only after decades of effort (for some examples, see Professor Robin Le Poidevin's wonderful book cited in Further Reading). A discovery made by medieval logicians may have a bearing on the existence of parallel universes. It centres on a result in logic known as Ex Falso Quodlibet, according to which the existence of just one contradiction logically implies that anything and everything is true. This has long been interpreted as a demonstration of the lethal effect on rational argument of contradictions, whose potency has terrified philosophers since the time of Aristotle. Yet if we do indeed exist in a multiverse, there aren't just a few contradictions out there, but an infinite number. Fortunately, we don't come across them because of the sheer size of the multiverse, which makes the chances of even one contradiction appearing in our tiny corner of it vanishingly small. Seen in this light, the supposedly absurd result that contradictions imply the truth of everything is an almost trivial statement about the multiverse – in which all things, including all contradictions, are true.

2. Physicists are mainly very pragmatic types who don't see the need to get bogged down in messy and subtle details unless it's absolutely necessary. Until recently, the interpretation of Quantum Theory was seen as the apotheosis of messy and subtle detail. But the emergence of specific applications such as quantum computation has compelled growing numbers of physicists to get to grips with these details. The result has been a revival in interest in the interpretation of Quantum Theory – and the emergence of a consensus that many of the standard textbook approaches taught for decades need to be revised. In 1999, attendees of a conference on quantum computation at the Isaac Newton Institute in Cambridge were polled for their views on how best to think of Quantum Theory. Out of the ninety who expressed views, only 4 per cent still adopted the standard ("Copenhagen") interpretation, while 33 per cent had adopted the Many Worlds interpretation. Over 50 per cent, however, described their view as being "none of the above" or undecided. For an accessible online review of the current status of interpretations of quantum theory, see http://tinyurl.com/8ycq9.

Further reading

Quantum: A Guide for the Perplexed by Jim Al-Khalili (Weidenfeld & Nicolson, 2003)

The Fabric of Reality by David Deutsche (Penguin, 1997)

Travels in Four Dimensions by Robin Le Poidevin (Oxford University Press, 2004)

25
The Anthropic Principle

IN A NUTSHELL

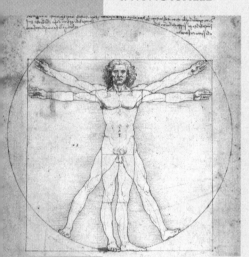

The immensity of the universe makes the notion of a connection with life on Earth seem hopelessly naïve. Yet discoveries about the properties of the cosmos suggest that such a link does exist, and has led to the Anthropic Principle: that our universe has properties that are suited for life – because otherwise we would not exist. From the discovery of special properties of carbon atoms needed to produce life-giving chemicals in stars to extraordinary "fine-tuning" in the properties of the universe, there is now a lot of scientific support for the Anthropic Principle. The growing suspicion is that our universe is just a small but life-friendly part of a far larger multiverse, where conditions just happen to be right for human life. Some scientists are now studying the possibility of a link between the existence of humans and the origin of the universe itself.

Albert Einstein called it the question he most wanted to answer – the ultimate "why" question: why is the universe like it is?

Some are certain they know the answer. They insist that the universe is like it is because God wanted it that way. For many others, this begs more questions than it solves, while some argue that the question is simply meaningless: the universe just is.

Yet recent discoveries about the cosmos suggest the truth is much more interesting, and that its properties – perhaps its very existence – are intimately linked to our presence within it.

The evidence takes the form of strange coincidences between basic properties of our universe and conditions needed to sustain intelligent life within it. The significance of these coincidences has prompted sometimes bitter arguments among scientists. Some dismiss attempts to make sense of them as unscientific. Others go as far as to insist that the links hint at divine meaning and purpose in the universe.

At the heart of the debate lies the Anthropic Principle, according to which the mere fact we exist casts light on what the universe can be like. As such, it has become one of the most controversial proposals in the history of science. This says at least as much about scientists as it does about the idea itself, for the notion that the

TIMELINE

1691   English scientist Robert Boyle argues that natural world appears to be specifically designed to support and benefit living organisms.

1903   British naturalist and pioneer of evolution theory Alfred Wallace argues that the existence of life demands that the universe be vast and complex.

1929   American astronomer Edwin Hubble makes first measurements of the properties of the entire universe, including its expansion rate.

1937   The Nobel Prize winning English physicist Paul Dirac reveals the mysterious Large Number Coincidences, claiming they imply gravity is getting weaker.

1953   The Cambridge astrophysicist Fred Hoyle uses the existence of life to predict the existence of a previously undetected energy level in an isotope of carbon.

1957   Princeton University physicist Robert Dicke shows that the Large Number Coincidences may simply reflect conditions needed for life to exist today.

1969   Apollo astronauts deploy laser mirrors to measure the Moon's distance, which reveal that gravity is not weakening – contrary to Dirac's theory.

1973   Cambridge astrophysicist Brandon Carter first coins the term Anthropic Principle for argument that properties of the universe are linked to existence of life.

1981   Alan Guth of the Massachusetts Institute of Technology puts forward inflation theory to explain the propulsive energy of the Big Bang.

1983   Russian cosmologist Andrei Linde at Stanford University puts forward chaotic inflation theory, producing lots of universes – some of which inevitably support life.

1992   American physicist Lee Smolin publishes theory suggesting universes undergo Darwinian-style evolution that makes them suitable for the existence of life.

universe is tailor-made for humankind was taken as read by many of the founders of western science, including Isaac Newton. But by the start of the twentieth century, such beliefs had acquired an air of religious dogma which most scientists were keen to avoid. By 1903, even the distinguished British naturalist Alfred Russel Wallace[1] risked ridicule by declaring that the universe may have the properties it does "in order to produce a world that should be precisely adapted in every detail for the orderly development of life, culminating in Man".[2]

Within a few years, Albert Einstein had developed his theory of relativity, marking the start of modern science, which seemingly had little use for such sentiments. By 1930, astronomers had found that the universe is utterly unlike the simple, static infinite void envisaged by earlier scientists. The universe was dynamic, teeming with galaxies all racing away from each other in the aftermath of an explosive event that seemed to have taken place billions of years ago. Its behaviour could only be understood in terms of the fearsomely complex equations of General Relativity, Einstein's theory of gravity. Certainly there seemed no place in this emerging view of the universe for a connection with the existence of mere humans.

Even so, some scientists claimed there was something odd about the properties of the universe, which hinted at an unlikely link with the physics of sub-atomic particles. They pointed out that according to the astronomical data, the visible universe was around 10,000 billion billion billion billion (10^{40}) times bigger than an electron – a huge number, but one which just happened to be the same factor by which the strength of the electrostatic force between electrons and protons exceeds the gravitational force between them. Was the similarity of these two large numbers just a coincidence – or was it evidence of a link between the physics of the very large and the very small?

While many scientists gave this very short shrift, it attracted the attention of one of the founders of Quantum Theory, the Nobel Prize winning British physicist Paul Dirac. In 1937, he published a paper in the influential journal *Nature* putting forward what became known as the Large Numbers Hypothesis. According to this, the similarity of the huge numbers is no coincidence, but reflects a fundamental law of physics.[3] As such, the similarity must always hold true – which led Dirac to make an astonishing prediction. Because of the expansion of the cosmos, the Large Number based on the size of the universe changes over time. As a result, its current similarity to the second Large Number cannot hold true forever – unless, that is, the second Large Number also changes with time. Yet what sub-atomic particle could possibly vary? Dirac decided there was only one way to avoid wrecking Quantum Theory: the strength of gravity must be weakening over time, at the same rate as the universe expands.

Dirac's daring proposal provoked relatively little immediate response – not least because the predicted decline in the strength of gravity was incredibly slow – amounting to less than 1 per cent over 10 million years. In the absence of any obvious test of this prediction, scientists put the Large Numbers Hypothesis on the shelf. The possibility of a link between the physics of the very large and very small

JARGON BUSTER

Anthropic Principle: The idea that the nature of our universe is intimately linked to the existence of observers within it – namely, humans.

Weak Anthropic Principle: A relatively modest version of the Anthropic Principle, first identified by the Cambridge physicist Brandon Carter, and the one most scientists are willing to accept. According to the WAP, the fact that at least part of the universe contains observers puts constraints on what the whole universe can be like.

Strong Anthropic Principle: A far more controversial version of the Anthropic Principle, which states that the universe must have properties which allow the emergence of life at some time during its history. To many scientists, this comes perilously close to being a statement of religious belief.

Large Numbers Hypothesis: The idea that the similarity of large numbers – typically around 10^{40} – that emerge from certain combinations of basic properties of the universe is no coincidence, but reflects a deep connection between cosmic and sub-atomic physics. Once advocated by such luminaries as the British Nobel-Prizewinning physicist Paul Dirac, the hypothesis has now been largely abandoned.

Inflation Theory: The leading contender for explaining the explosive force of the Big Bang 14 billion years ago, which attributes the rapid expansion to so-called quantum vacuum energy. Emerging literally out of nowhere, this energy acts like a form of anti-gravity, "inflating" the universe up from its original sub-atomic size.

Chaotic inflation: A version of Inflation Theory developed by the Russian cosmologist Andrei Linde, which leads to the creation of lots of separate regions of the universe, only some of which may support life.

The electrostatic force

One of the fundamental forces of the universe, the electrostatic force binds together oppositely-charged objects – such as protons and electrons – and is thus responsible for the existence of atoms. It also ensures that like charges repel, and plays a key role in nuclear reactions inside stars, which can only start once the natural repulsion of atomic nuclei for each other has been overcome. Even a slight change in the strength of the electrostatic force would therefore have a profound effect on the nature of our universe, preventing certain types of star from existing. That, in turn, would lead to a lack of key chemical elements normally created inside the core of giant stars – and rule out the ingredients for the formation of intelligent life such as humans.

burst back onto the scientific scene in 1953, when Dirac's Cambridge colleague Fred Hoyle uncovered a far more impressive "coincidence" – one which gave a direct link between the properties of the cosmos and the existence of life on Earth.

Since the mid-1940s, Hoyle had been wrestling with one of the major challenges confronting science: explaining the origin of the chemical elements. The simplest and most common of these – hydrogen and helium – seemed to have been created in the unimaginable heat of the Big Bang. The problem was to explain the origin of all the others.

Hoyle found that the answer lay in nuclear reactions taking place deep inside stars – but only if there was something very special about atoms of carbon. In particular, one of its isotopes – "C-12" – had to possess a so-called resonance at a very precise energy. If it didn't, the stars would create virtually no carbon – and, as Hoyle realised, no carbon meant no life. Yet try as he might, Hoyle could find no experimental evidence that such a resonance existed. Instead, he resorted to a characteristically ingenious argument: the experts must

have missed something, because if they hadn't, they would simply not exist.

Hoyle's pioneering use of the Anthropic Principle inevitably provoked scepticism: surely the mere existence of life could not be used to predict a fundamental property of atoms? Yet Hoyle persisted, and barely a week later, a team of experimentalists came back with the news that they had found the resonance – right where Hoyle said it would be.

The Anthropic Principle received a further boost shortly afterwards, when the American physicist Robert Dicke used it as the basis for a radically different way of thinking about Dirac's Large Numbers Hypothesis. Dicke revealed a loophole in Dirac's argument that the similarity in such huge numbers could hardly be a coincidence. He pointed out that the fact that we are able to observe the "coincidences" means that the universe must have existed long enough for stars to create elements needed for human life. On the other hand, the universe cannot be much older than this, otherwise the raw materials needed to create stars would have been used up. Using a simple model of stellar nuclear reactions, Dicke showed that our existence ensures that Dirac's two Large Numbers must be roughly similar at this point in the life of the universe. In other words, Dirac was right that their similarity is not just a coincidence – but wrong to think it must therefore persist forever.

Dicke's anthropic argument thus also undermined Dirac's claim that gravity must be getting weaker to keep the "coincidence" intact. Doubts about Dirac's prediction had already surfaced in the 1940s; the Hungarian-American theorist Edward Teller had shown that weakening gravity implied a brighter sun and smaller Earth orbit in the past, making the primordial Earth impossibly hot.[4] The *coup de grâce* was delivered by the Apollo astronauts, who in 1969 left special laser-reflecting mirrors on the Moon. These allowed astronomers to measure the distance to the Moon to an unprecedented accuracy of just a few centimetres – and thus detect any signs of it drifting away as gravity loosened its grip. By the 1990s, the conclusion was clear: if gravity is weakening at all, it does so far more slowly than predicted by Dirac's theory.[5]

Despite the growing interest in the anthropic principle, it was not until 1973 that it was finally given a name, when the Cambridge University astrophysicist Brandon Carter coined the term from the Greek word "anthropos", meaning mankind. Carter also distinguished two varieties of the basic idea. The "weak" Anthropic Principle – the variety used by Hoyle and Dicke – is relatively uncontroversial, and states that the fact we exist puts limits on certain properties of the universe and its contents. The "strong" Anthropic Principle, in contrast, asserts that the universe is actually compelled to have properties compatible with intelligent life. For many scientists, this smacks of the notorious "Argument from Design" developed by Victorian theologians, according to which the miracles of nature prove the existence of a supreme designer – namely, God.[6]

While few scientists accept the strong version of the Anthropic Principle, even the weak variety has come in for criticism. Sceptics argue that it simply reflects what we know to be true – that we exist – and is thus devoid of content. For some scientists

Fred Hoyle

Unquestionably the most brilliant astrophysicist of the twentieth century, Fred Hoyle revolutionised understanding of both the universe and its contents. In 1948, he put forward the Steady State theory, which argued that the cosmos has always existed, and remains unchanging on the grandest scales. While disproved by the discovery of the heat left over from the Big Bang, the Steady State theory prompted many theoretical and observational advances. Hoyle also solved the problem of the origin of the chemical elements in stars, and was scandalously robbed of a Nobel Prize following his past criticism of the prize-giving committee. In later life, Hoyle espoused many unpopular theories, including the idea that life came to Earth as microbes carried aboard comets.

– including Sir Martin Rees, the present Astronomer Royal – this goes too far the other way. Instead, they have pioneered a middle way, which leads to one of the most spectacular of anthropic conclusions: that our universe is merely a tiny but habitable part of a truly enormous "multiverse".

On this view, asking why our universe happens to be "fine-tuned" for life is like asking why card players sometimes get a Royal Straight Flush: it's bound to happen somewhere, sometime in the multiverse.[7]

The possibility that our universe is just a tiny part of something far larger is backed by theories developed to explain the Big Bang. In 1981, the American physicist Alan Guth showed that the cosmic expansion may have been triggered by a sub-atomic force that emerges literally out of empty space, and acts like a form of anti-gravity. Two years later, the Russian cosmologist Andrei Linde showed that this "inflation" effect could have produced a vast variety of universes, only some of which – like the one we inhabit – are suitable for life.

In 1992, the American theoretical physicist Lee Smolin took this idea a step further, proposing that all universes take part in a kind of Darwinian "survival of the fittest", which produce conditions suitable for life as a by-product. Smolin argued that fresh universes can be formed via inflation taking place inside black holes, the superdense remnants of giant stars whose gravity is so strong that not even light can escape from them. Conditions in each of these new-born universes are slightly different, but those able to produce the most black holes will also produce the most offspring – and so become the most common form of universe. And this leads to a possible link with the existence of life – for as Smolin points out, universes that produce the most black holes will also produce the most giant stars, which are just the ones best able to create the chemical elements needed for life.

Though highly speculative, Smolin's proposal is broadly in line with known laws of physics, and may one day even be testable. If confirmed, it would give a whole new perspective on the Anthropic Principle, by tying our existence in the cosmos to the existence of the cosmos itself.

That, in turn, leads to the most astonishing implication of the Anthropic Principle, and one that will keep scientists arguing about it for years to come: that we need the universe to exist – and it needs us.

Notes

1. One of the most brilliant Victorian naturalists, Wallace is widely regarded as co-discoverer of the principles of evolution: his letter to Charles Darwin about his findings in 1858 spurred the publication of *The Origin of Species*. In later life, Wallace became interested in more spiritual aspects of biology – which may have led to his prescient writings about links between the existence of life and the nature of the universe.

2. According to the Anthropic Principle, the universe we see has properties consistent with the existence of humans – because if it didn't, we wouldn't be here to tell. In 1980, the American physicist William Press came up with an ingenious twist on this argument which predicts the maximum possible height for human-like creatures in our universe.

   The argument centres on the fact that two-legged creatures like humans are perpetually threatened by injury from falling over. Clearly, if we were much taller, we would strike the ground with so much energy that death would always result. Press pointed out that the reason this doesn't happen ultimately lies in the relative strengths of two fundamental forces in the universe: gravity – which dictates the energy with which we hit the ground – and the electromagnetic force, which governs the strength of the atomic bonds making up our bones.

   This argument leads to a formula for the maximum safe height for humans in terms of the strengths of these two forces in our universe, and gives a figure of around 3 metres. As it happens, the tallest human who has ever lived – Robert Wadlow (1918–1940) – was within this limit, at 2.72 m.

3. The Nobel Prize winning physicist Paul Dirac believed that the similarity of certain special numbers hinted at a link between sub-atomic physics and the universe. While his original theory is now known to be wrong, the idea of

such a cosmic connection has now become standard thinking. The expansion of the entire universe is thought to be propelled by a force generated by quantum processes – the laws governing the sub-atomic world. So far, theorists have been unable to calculate the strength of this cosmic force using standard methods. Intriguingly, however, it is possible to use Dirac's special "large numbers" to come up with an estimate – and the result is close to that measured by astronomers. So perhaps Dirac's crazy idea was on the right track after all.

4. In Dirac's original theory, the link between sub-atomic physics and the cosmos led to an astonishing prediction: that gravity must be getting weaker. It took some time for physicists to work out the consequences of this prediction, and they proved to be disastrous for the theory. If gravity is getting weaker, then the attraction of the sun on the Earth would have been stronger in the past, pulling it closer in – and thus making it hotter. In addition, the change in the strength of gravity would alter the balance of the forces inside the sun that drive its nuclear fusion reactions. Calculations show that just a 5 per cent increase in the strength of gravity would drive up the luminosity of the sun by 40 per cent. Combined with the smaller orbit, the resulting temperatures on Earth would exceed that of boiling water just 600 million years ago – in flat contradiction of the fossil record.

5. Among the experiments deployed during the first Moon mission in July 1969 was a 77-kg array of 100 highly polished mirrors. Designed by Dr Carroll Alley of the University of Maryland, these reflect back laser beams sent out from Earth, and the time taken for the round trip combined with the speed of light reveals the Earth-Moon distance to a precision of 1 part in 10 billion, or about 3 cm.

6. Widely used by Creationists in debates against Darwin's theory of evolution, the Argument from Design is usually attributed to seventeenth-century English theologian William Paley, who claimed that the existence of exquisitely complex living organisms is best explained as the result of deliberate actions by a Creator. Darwinian scientists insist that evolution can account for the wonders of life without a Creator.

7. A striking example of apparent cosmic "design" centres on a number called Omega, which gives the ratio of the actual density of matter in the universe to that needed for gravity to be strong enough to halt the cosmic expansion. Current observations show that Omega is around 0.3. Calculations reveal, however, that this means the original value of Omega at the Big Bang must have been within 1 part in a million billion of exactly 1.0. If not, the cosmic expansion would have been too fast for galaxies to form, or so slow that the universe would have collapsed billions of years ago. In short, we owe our presence to some astonishing "fine-tuning".

Further reading

The Anthropic Cosmological Principle by John Barrow and Frank Tipler (Oxford University Press, 1988)

The Life of the Universe by Lee Smolin (Phoenix, 1997)

Just Six Numbers: The deep forces that shape the universe by Sir Martin Rees (Weidenfeld & Nicolson, 1999)

The Accidental Universe by Paul Davies (Cambridge University Press, 1982)

Where Next?

The contents of this book are up to date at the time of writing, but science does not stand still. Those wanting to keep abreast of the latest developments as they happen have never been better served. There are many first-rate magazines, such as *BBC Focus* (in which shorter versions of some of these chapters originally appeared), *Scientific American and New Scientist*. For those who really can't wait for the latest breakthroughs, there are also many websites reporting the very latest developments, such as www.focusmag.co.uk, www.sciam.com and www.newscientist.com.

Whatever route you choose to follow the latest developments in science, I hope you will find this book a handy companion for making sense of their implications.

Index